CAMCORDER
KNOW HOW

MARSHALL CAVENDISH

This edition published in 1993
by Marshall Cavendish Books
(A division of Marshall Cavendish Partworks Ltd)
119 Wardour Street
London W1V 3TD

ISBN 1 85435 719 0

Material previously published in the Marshall
Cavendish partwork *CAMCORDER KNOW HOW*

Printed in Great Britain

FOREWORD

Camcorder Know How, written by a panel of camcorder experts, provides you with all the information you need to improve the quality and content of your home videos. From making lively recordings of family events to assessing and editing your existing tapes, you will find the relevant know-how clearly set out and explained. Photographic sequences with special shooting guide symbols, text panels and diagrams combine to make *Camcorder Know How* the answer to all your videomaking problems, large or small. The book covers six main areas:

VIDEO TECHNIQUES

A step-by-step guide to improving your skills: Opening shots · Zoom in! · Sound recording · Sharp focusing · Steady as you go! · Colour control

What marks out the knowledgeable camcorder user is a grasp of techniques borrowed from the professionals. Even the simplest of these will greatly help your video tell its own story and enhance the enjoyment of your audience. *Camcorder Know How* introduces you to six techniques. Each is accompanied by a simple project so that you can practise your new skill easily and add it to your growing repertoire.

PRACTICAL EDITING

Tips and hints from the professionals: Think before you shoot · The editor's eye · Taking the lead · Assemble editing · Editing for creative impact · Dubbing music

Using simple editing skills can make all the difference to your finished video by adding extra drama or changing your recording's pace. Editing can literally bring your recordings to life. *Camcorder Know How* guides you through a range of techniques, from simple point-and-shoot methods while you are recording to the more complex processes involved in editing after the event.

ON THE DAY

How to record special occasions: Happy Birthday! · On safari! · The big match · Down on the farm · The sky's the limit! · Beside the seaside

The most treasured videos are those recording important family occasions, such as birthdays, holidays and days out. *Camcorder Know How* guides you through all the special considerations you will need to take into account so that you can be guaranteed to get the best from these fun and often emotion-packed days.

EQUIPMENT FILE

An in-depth look at accessories: Tripods · Tapes unravelled · Batteries · Hand-held microphones · Camcorder carriers · Battery pack lamp

Whatever camcorder you own, there is a whole range of accessories available to help you get better sound or pictures or take care of your equipment. If you are working to a tight budget, it is vital to know what to look for when you buy. *Camcorder Know How* goes behind the sales hype to reveal how products really work.

VIDEO MATTERS

Camcorder talking points: Camcorder care · Music and copyright · Camcorder travel · Your videos on TV news · A world of difference · Anti-shake systems

Camcorder Know How explores a range of topics which commonly intrigue or puzzle the camcorder user.

VIDEO TERMS

An illustrated guide to video terminology.

Knowing the right terms for what you are doing is more than a matter of pride – it can help you solve a technical problem or get the very best from your videomaking. *Camcorder Know How* guides you through the range of terms from Aberration to Autofocus.

Opening shots

PHOTOGRAPHS BY STEVE LYNE

FRAMING THE RIGHT SHOT AT THE RIGHT TIME IS ONE OF THE SECRETS OF A WELL-MADE VIDEO. ALL YOU NEED FOR GOOD RESULTS IS TO FOLLOW A FEW BASIC RULES

One of the most obvious differences between an experienced camcorder operator and someone who doesn't know their white balance from their fader is their ability to frame a shot. Unfortunately, no matter what gadgets your camcorder possesses, there is a good deal more to framing a shot effectively than just pointing and shooting.

Few things are more guaranteed to irritate anyone watching a video (and embarrass the person who has created it) than sequence after sequence featuring people with their heads cut off by the frame, people 'falling out' of shot, distracting

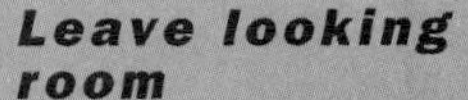

LEARN HOW TO ▶ ***Compose the basic shots*** ▶ ***Leave looking room*** ▶ ***Allow for movement in the frame*** ▶ ***Apply the Rule of Thirds***

LONG SHOT ▶
Most sequences start with a long shot that places the subject in context.

◀ MID SHOT
This staple shot of television drama focuses attention more closely on the subject while retaining some background detail. The cut-off point should be just below the waist, with some room left above the head.

CLOSE-UP ▶
This head-and-shoulders shot is ideal for showing the subject's expression and mood. The cut-off point should be just below the shoulder.

objects in the background and crucial moments of action happening off-camera. But with just a little practice using basic framing techniques, bloomers like these can be avoided and your enjoyment of your camcorder hugely increased.

VIDEO LANGUAGE

A good shot, carefully composed, draws the viewer into the story your programme will unfold. There are three basic shots in filming: the Long Shot, the Mid Shot and the Close-Up.

The long shot (sometimes called a wide shot), includes the whole picture and relates the subject to his or her environment. Pull back further and it becomes a very long shot or an extreme long shot. Long shots are very useful as 'establishing' shots, to give a broad view of your location and introduce your subject.

The mid shot should extend to just below the waist of the subject. (Whenever two people are shown, it is called a two-shot.) Mid shots concentrate on the subject but also include some background details.

MOVING CLOSER

A close-up is a head-and-shoulders shot of a single person. When only the face occupies the frame, the shot becomes a big close-up. When just eyes and/or mouth are featured, you have an extreme close-up.

Each basic shot has a different emotional impact on the viewer. For example, an extreme close-up is a very intimate view of a person; they bring us closer to a subject than we would normally be in real life. On the other hand, someone talking in long shot will appear distanced and isolated (their words are likely to be obscured by other background noises, too).

When filming a person (for practice, find someone who will not mind taking a bit of direction and who will keep still for you), a long shot can be a good starting point as it

▲ BIG CLOSE-UP
This removes all the background and focuses on the subject's eyes and smile.

▶ EXTREME CLOSE-UP
This shot should only be used for dramatic effect. Not everyone will appreciate being looked at so closely.

is an ideal way to show gestures and body movement and places your friend in some sort of context.

Ask them to stand 5 to 8 metres away from you, in front of a pleasant background. Frame up a long shot, being careful to leave a little space above the top of their head. This is called headroom; too little and you cut off the top of their head, too much and they look as if they have sunk to the bottom of the frame.

A NEW ANGLE

After this, try moving to a mid-shot – but take it from a slightly different angle. This will increase viewers' interest in what you are recording. You can then move slowly in for a close-up, featuring your friend's head (allow for a little space above the head) and shoulders.

End the sequence with an extreme close-up, in which your friend's face fills the frame from top to bottom.

Whenever framing a shot, you need to bear in mind an artistic rule, which applies to painting and photography as well as video: the Rule of Thirds. This principle dictates that the main focus of interest in a picture lies approximately a third of

CAMERA SENSE

LOOKING ROOM

When composing a shot from the side, leave plenty of space in the frame for your subject to look into (right). This shot logically implies that your next shot will show what your subject is looking at. The poorly composed shot (below right) gives the impression that the subject is staring at a blank wall, ruining what could be an interesting sequence. This framing error also leaves a large amount of 'dead' space behind the subject's head.

the way in. In a well framed close-up, for example, the subject's eyes should appear about one third in from the top of the frame.

Of course, people – especially children – often have big problems taking direction and keeping still, so strictly following the rule of thirds can be difficult. But if the horizon (which doesn't jump around) features in your picture, make sure that it is either a third or two thirds in from the top of the frame. Another important tip is always to ensure your camcorder is level and that straight lines – such as the horizon or buildings are parallel to the edges of your viewfinder.

It's best to avoid putting important subjects right in the centre of a shot, as this tends to look formal and uninteresting. Be especially careful not to let the edge of the frame cut through important action, or cut off bits of people. Either put them in the frame, or right out of it. It is also a good idea to leave a generous space around the edge of your shot as some televisions may not show as much on screen as the viewfinder of your camcorder.

'Epic' scale shots, so effective on a cinema screen, never look as good on video, with people in them reduced to tiny blobs. Go instead for closer shots of interesting reactions and details, put together into sequences that 'tell a story'.

Q

What is the best way to video my sons playing cricket in the garden?

Don't try to follow the action or the results will look very wobbly, with your sons whizzing in and out of the frame, making the action impossible to follow. Take up a camera position that takes in as much of the playing area as possible and let the action happen for the camera.

A

IMPROVE YOUR SKILLS

A Walk in the Park

Framing people is a simple matter when they keep still, but composing good shots in which your subjects are moving takes a bit more practice. A good way to improve your ability to compose for moving action is to ask a couple of friends to be your models and practice framing them while they take a stroll.

▲ SCENE SETTING Begin your sequence with a long shot, lasting a few seconds, of the park with your subjects moving on a diagonal from left to right to give them maximum room to walk into.

▶ LONG SHOT Keeping the camcorder steady, move in, with your subjects still in long shot. As they exit from the frame, press Pause.

▶ HEAD-ON SHOT
Change your position, keeping subjects in long shot, to follow on from the previous sequence. Then allow them to move towards you for a mid shot. Press Pause.

▶ TWO SHOT
To end, ask your friends to sit on a bench and frame a mid shot. Then zoom in for a close-up, in which they turn and face the lens.

If shooting people from the side, always allow them plenty of 'looking room' in the frame. If a person is looking to your right, position them on the left of the frame. Try to avoid leaving too much space above your subject's head or they will appear to be in the process of slipping off the bottom of the screen.

When framing shots of people, take care not to come in too tightly around their heads, waists, knees or ankles, the natural joints of the body, or they will appear 'cut off' from the rest of their bodies. When framing a head shot, remember to leave a little of your subject's shoulder in the frame. The viewer's mind's eye will fill in the rest of their body. The same applies with mid shots: frame your subject just below the waist.

FUNNY HATS AND FALLING STARS

Try to be aware of everything in your viewfinder, the background as well as the foreground. But be careful the background does not distract from the action, or you could be in for a bit of a shock when you play your tape back on TV. A telegraph pole positioned a good 30 metres away will seem to be growing out of your subject's head if they stand in front of it, or a bush in the distance may

ASPECT RATIO

Camcorders, unlike still cameras, cannot be turned on their side to give a portrait-style shot, so you can never change the shape of your picture. The image you put on screen will always be four broad by three high. This international standard measurement is known in the profession as Aspect Ratio.

CAMERA SENSE

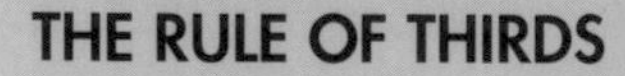

THE RULE OF THIRDS

This rule is an invaluable guide to the use of screen space. Imagine a noughts and crosses grid placed over your shot. Try to put points of visual interest on or near the places where the lines intersect: the eye is naturally drawn to these areas. If you are taking a shot featuring two people, make sure one person is one third of the way into the frame from the left, and the other is one third in from the right. In a close-up of a face, compose your shot so that the eyes are placed one third of the way from the top of the frame. In composing scenic shots, the rule ensures that the horizon and verticals are balanced.

blend in with their hair or body to unintentionally comic effect.

There will, of course, be many occasions when you will want to video people walking along or running about. In every case, you should allow plenty of room in the frame for them to move into. Position yourself at an angle to your subject to create the effect of them moving in a diagonal line across the frame. This technique allows him or her maximum room to walk into and prevents your subject appearing to 'fall out' of the frame at the side as they pass (a common fault in beginners' videos).

For example, if you compose your shot so that your subject enters the frame at the left hand top corner of the screen, they should exit near the bottom right hand corner (the direction of the movement does not much matter, so long as the subject enters and leaves near the corners of the frame).

For head-on shots you should place your subject right in the middle of the frame. Then, as your subject comes nearer and nearer to where you are standing , move slowly to one side to allow them room to exit at a corner of the frame (and not at one of the sides, which looks ugly).

If these basic techniques are employed every time you use your camcorder, you will be amazed at the improvement in your videos. With some practice, framing will become second nature – and your films will steadily acquire a polished, professional look. □

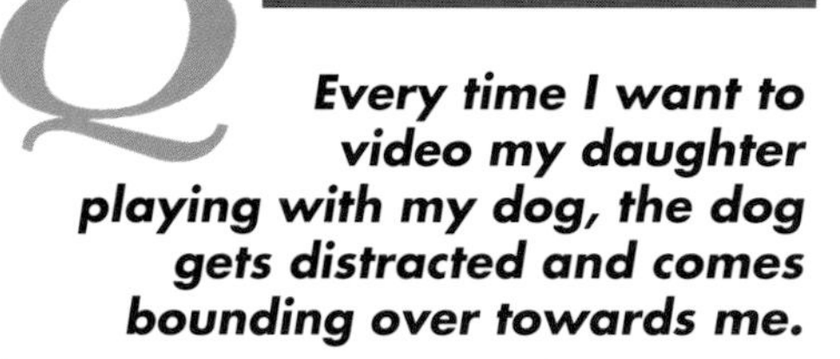

Q ***Every time I want to video my daughter playing with my dog, the dog gets distracted and comes bounding over towards me.***

A *The secret is not to get too close. Stand well back zoom in on them both. Use a tripod if necessary to avoid camera shake.*

OUT-TAKES!

A BUNCH OF BLOOMERS!
It's surprisingly easy to make a framing error – especially when people are moving around.

• **GOING NOWHERE** – THIS COUPLE HAS BEEN FRAMED TO EXIT FROM THE SIDE, INSTEAD OF AT THE CORNER

• **SEVERED HEAD** – THIS CLOSE-UP SHOULD INCLUDE SOME OF THE GIRL'S SHOULDER TO PREVENT HER IMAGE SEEMING TO SLIP OFF THE BOTTOM OF THE FRAME.

• **ME AND MY SHADOW** – SHOOTING WITH THE SUN DIRECTLY BEHIND YOU CAN HAVE ITS PROBLEMS . . .

Think before you shoot

YOU DON'T NEED TO SPEND HOURS FIDDLING WITH SOCKETS, PLUGS AND WIRES IN ORDER TO EDIT YOUR VIDEOS — WITH A LITTLE FORETHOUGHT AND A FEW BASIC SKILLS, YOU CAN EDIT AS YOU GO!

One of the main skills you will need to develop for making home videos is editing. Editing essentially means putting the shots of your video in the order and form in which you want them to be seen. It is only through editing that your recordings take on the identity of a film.

There are two basic ways in which you can edit your recordings. One, the more complex, is the assemble method, where you use your camcorder, TV, and VCR (or, if you have one, editing controller) to re-record sequences on to a new tape in the order you wish — cutting out the shots you don't want and re-ordering the ones you do into their correct sequence. The other method is simpler, and you don't need to use any equipment beyond the camcorder itself. This is called in-camera editing, because the film is 'edited' as you shoot.

Editing in-camera is an essential skill. Since each sequence will be viewed exactly as it was shot, both before you record and as you shoot, you are forced to consider which events you really want, what order you want the shots to run in, the ways in which you can take the shots, and how fast you want the film to move. Even camcorder users with hi-tech editing suites will find it useful, for once you have mastered the art of in-camera editing you will have far less superfluous footage to cut out, and even short sequences will have a more polished appearance.

In-camera editing requires some advance planning and a good deal of on-the-spot

ILLUSTRATIONS BY EMMA WHITING

▲ To capture the highlights of any event, your finger should be constantly ready on the record button. During a football match, pause during the least exciting periods of the game, then start to record when the action appears to pick up.

SOUND ADVICE

THE ART OF NOISE

The sound-track of your video is as important as the images themselves. So unless you intend to dub music over your sequences, you will have to practice thinking in terms of sound as well as pictures. Cutting from a noisy street to a quiet waterway may help to build up the atmosphere of your film. Yet if you cut back and forth continually from the noisy scene to the quiet one, it will make the video distracting to watch.

shrewdness. Every event has key moments that you will need to capture to tell a coherent story. With many of these, there is no second chance — you cannot un-cut a wedding cake or ask a go-karting champion to repeat his winning lap; so you want to be in the perfect spot to capture the moment in the best possible way.

The first essential is some knowledge of the subject you are going to film. You may want to record the streets of the picturesque country village where you are spending a weekend, in which case take a wander round before you start and make a note of what you want to shoot, from which angle, and how. Think how the shots will link as a sequence. For recordings such as this, where your subject is mostly static, you don't need to plan in detail; but the more thinking you do in advance, the better the end result.

ADVANCE THINKING

If the event is public, such as a village fete, it helps to know the order of events. There may be a printed programme available; try to obtain one in advance (even if it is only on arrival at the gate) so you can work out where you want to be and when — particularly if there are several interesting things happening at the same time.

But at first, it is likely that most of the events you record will be small, family affairs

PLANNING THE MATCH STEP-BY-STEP

1 The teams come out. A relatively long establishing shot of 10 seconds or more is the right option here.

6 At the end of the game, a lingering 15 second shot of the victorious side provides a fitting close to your film.

5 Take a risk and position yourself behind one of the goals to capture one of the game's climaxes — a goal.

without a fixed itinerary. In such cases, it is well worth jotting down your own mini-programme. If you want to record a barbecue, for instance, check the layout of the garden in advance, even if you think you know it off by heart. Note the places where children will be most likely to play, where the guests will stand or sit and talk, and so on. Then think of the shots that will best depict the event: the guests arriving, children playing, meat going on the barbecue, glasses raised in a toast, the guests leaving.

Shooting order

Remember that shots ought to have some sort of understandable sequence: it may look more effective if a shot of the guests drinking is preceded by one of the bottles being opened! In the same way, the shot of bottles being opened might be preceded by a shot of them being brought out into the garden. In other words, try and give your film a sense of the progression of the actual event.

Bear in mind that the pace of your film – how fast it seems to move – will be determined not only by the length, but by the

BACKSPACE EDITING

When you press the on-off record button to pause your recording, your camcorder automatically winds the tape backwards over the last second or so of the previous shot. When you press the button to start recording again, the tape runs forward to a point just before where you had stopped and then starts to record the new shot. You will thus lose a small amount at the beginning of each shot – and a little at the end. This may be quite noticeable, so it is important to adjust your shots accordingly. To avoid losing vital moments, press Record a couple of seconds before the action begins and leave a little extra at the end of every shot.

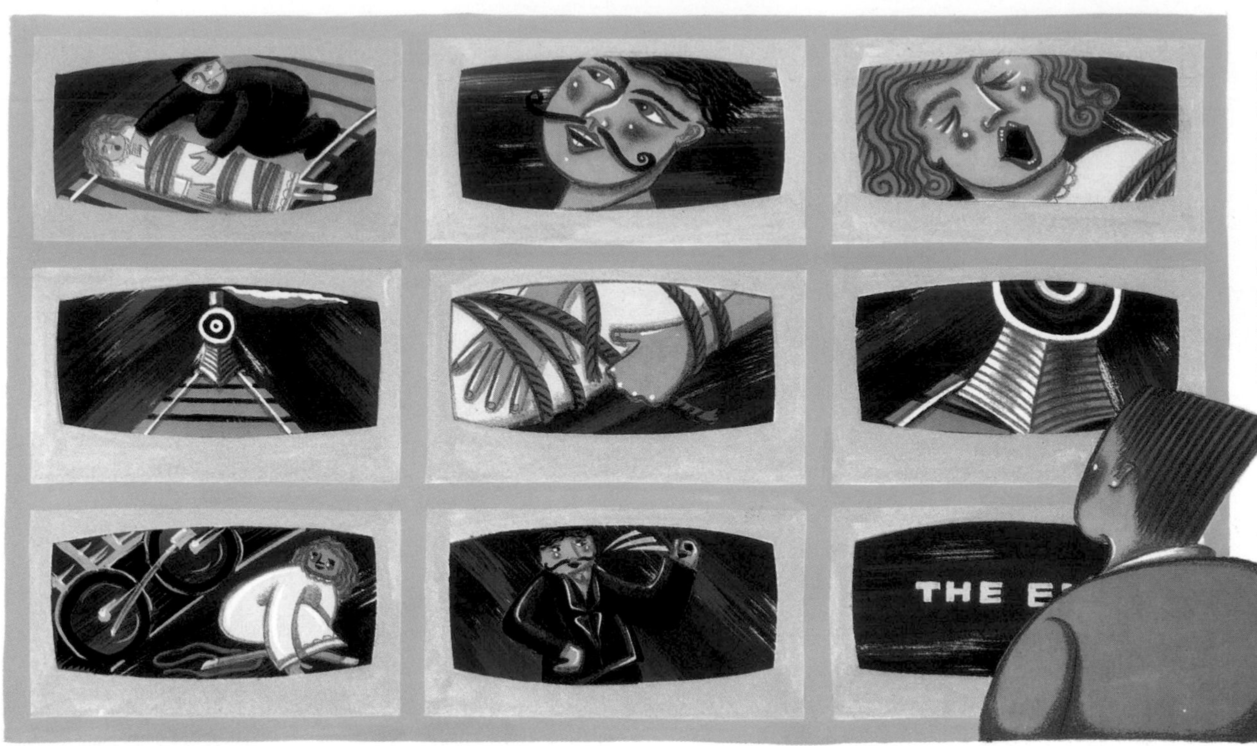

◀ **ACTION ON FILM**
It is well worth noting of the ways film-makers edit to inject excitement into a sequence. Constant variation of the framing, as well as the length and angle of each shot, plays with the audience's responses. Juxtaposing an extreme close -up of the woman's bound hands with a big close-up of the oncoming train, for example, is an obvious way of creating audience tension. If your VCR has a slow-motion facility, study the editing in any action sequence of a film: here editing truly comes into its own.

YOUR BEST SHOTS

The length, angle and framing of each shot can make or break your video. Here is a rough guide to follow:

- Static subjects should be held for around five to seven seconds, but this can vary according to your shot. A new establishing shot could be held for up to 10 seconds; a close-up of a subject already featured in the video might be held for only three to five – unless the subject is speaking.
- All shots of written material such as street signs should be held just long enough to read.
- Allow your subject to dictate the pace of your video – some swans swimming peacefully on a lake, for example, will demand more lingering shots than skiers whizzing past you down a slope!

interest value of each shot. A five-second static shot of food on a grill will seem to last longer than a five-second action shot of children playing. So, as you shoot, let the interest value of the event dictate the length of each shot. In addition, avoid stringing static shots together – it will make your videos monotonous to watch. Try and interweave static shots with ones in which there is movement, always varying the pace.

QUICK REACTIONS

To make your film still more enjoyable, you might try to work reaction shots into your sequences. Say a dish of food is brought out into the garden; you may want to show the response this elicits from the guests. Break off from filming the dish, and redirect the camera to show the guests' delighted faces: then go back to a shot of the dish.

These are some of the basics of in-camera editing; they will not only give shape and sense, they will give variety to your films. And for the viewer, variety means interest. Obviously, you cannot expect to become proficient immediately at in-camera editing; but by starting to think about it now, it will soon become a natural part of your video-making practice. □

▲ **Notes, sketches, or just simple lists of the scenes you are about to shoot will help you to edit your films in-camera. The golden rules are:**
- **Familiarize yourself with the filming area.**
- **Be aware of the probable order of events.**
- **Plan main shots in advance.**

Happy Birthday!

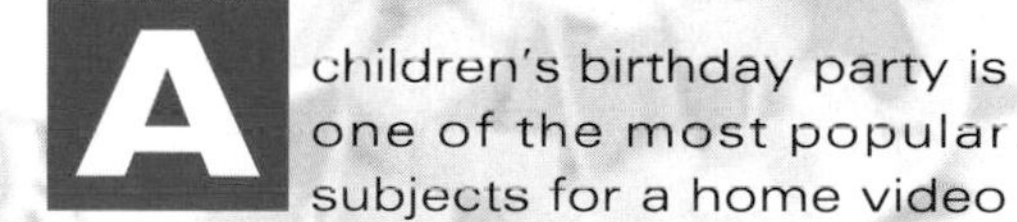

A PARTY IS THE HIGHLIGHT OF EVERY CHILD'S BIRTHDAY. CAPTURING THE EXCITEMENT ON VIDEO IS EASY – WITH A LITTLE PLANNING AND CARE

A children's birthday party is one of the most popular subjects for a home video – and one of the best, too. Unlike most adults, children are natural show-offs, especially when egged on by their excited friends. But unless you're perfectly happy with a crowd of eager little faces jostling in front of the lens every time you go for a shot, you may have to resort to a few cunning tricks . . .

As with every special event video, a degree of planning is vital. You will already have decided exactly which rooms in the house you wish to hold the party in – perhaps the lounge for games and other activities and the kitchen for the tea and cake ceremony. (One thing you don't want, especially if you're videoing the event for posterity, is to have screaming children running all over the house getting into mischief.)

PHOTOGRAPHS BY RAY MOLLER

Before everyone arrives, walk around these areas with your camcorder to check the lighting. Take a few test shots and play them back through your TV. You may need to switch on a few lamps to illuminate dark corners. You should also remember that, when shooting, you will have to keep your back to natural light coming in from windows, or you will cause your camcorder's auto-iris to close up, putting faces into deep shadow.

Lighting checks

A few practice shots will soon tell you where to position yourself to get the best of the available light. Try switching on overhead lights to see what effect they have. Now is the time to experiment, not later when the party is in full swing. It is very important that, during the event, you do not spoil the children's enjoyment by insisting that they try this or do that. It's their party, after all.

Rearrange furniture and the tea table to positions that will suit your needs – what you want is the best video coverage. You should also work out as many potential shooting angles in advance. Check how much of a room can be included with the lens at its widest position and see just how much you need to move the camcorder to include everything.

Place a chair in one corner to stand on for a high wide angle of the main party room. Later this will allow you to take in all or most of the action in a single shot instead of vainly trying to follow individual children charging about. Use this as a 'master' shot position that you can return to at various times during the progress of the party.

Edited highlights

As the camcorder operator, it is important that you keep in the background, observing the action. Don't expect to record every event.

PLANNING CHECKLIST

- **Decide on your best camera positions in the hall, main party room and around the tea table**
- **Decide on the best place for the conjurer to stand**
- **Check lighting levels in all of the rooms**
- **Practise shooting 'blind'**
- **Decide positions on other set-up shots, such as the Lucky Dip tub**

▲ Find a good place in the hall to record guests as they enter. Frame up a long shot. Mark position on floor. Play back to test lighting.

▲▶ In the main room, frame up a wide shot. Mark position; play back for lighting level as before.

▶ Light from windows may cause problems during tea. Shoot with the light behind you.

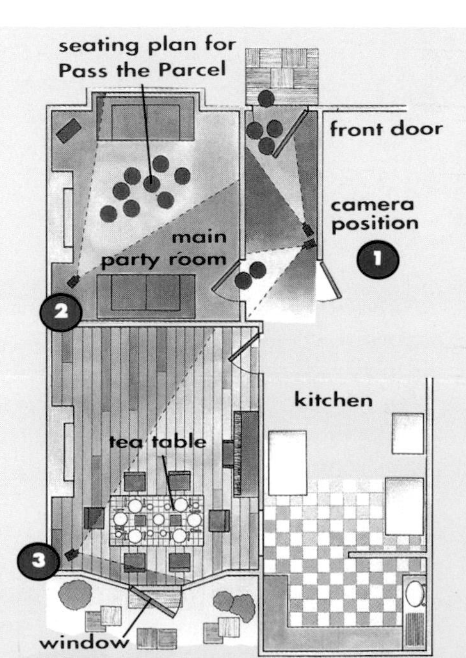

ILLUSTRATION BY BILL LE FEVER

PARTY PLAN

To remind yourself of the best places to make the most of the action, you may like to draw a rough plan of the party rooms. Indicate where you would like the children to sit for a party game such as Pass the Parcel. Remember to include the windows and also other light sources that you plan to use. Clear rooms of any unnecessary items of furniture.

PARTY TITLES

To give the beginning of your birthday video a touch of class, try shooting the following title sequence before the party starts. Line up birthday cards on a shelf with the party invitation at the end of the line. Check framing for a close-up of the cards. Ask a helper to put some music on the hi-fi and slowly move the lens along the line. When you reach the invitation at the end hold it for a few extra seconds.

◀ Right on time, the first guests arrive to be greeted by the hostess. Record from the long shot position in the hall (watch out for the light coming in from the door).

▲ Cut to a close-up of each guest, slowly pulling back as the group move towards the main party room.

Be selective and turn off when things get dull or disruptive. A finished tape of around 20 minutes will make a more entertaining memento than two hours of continuous recording.

TAKING CHARGE

Two or three other adults are essential to take charge of proceedings – to organize games and the tea, mop up spillages and jolly along any child overcome by excitement or distraught at coming second at Musical Chairs.

Choose the more visual games where the children are all together – Grandmother's Footsteps, Statues and Pin the Tail on the Donkey are ideal. Avoid hunting games which are difficult to follow or verbal games that require continuous sound.

A STORY TO TELL

The best home videos tell a story and so your party should have a definite order of events, which you as 'director' should be thoroughly aware of in advance. This might begin with an opening shot of a few balloons tied up outside your front door indicating where the party is, followed by a few shots of your children getting ready. At this stage include a wide shot of the laid party table and carefully composed shots

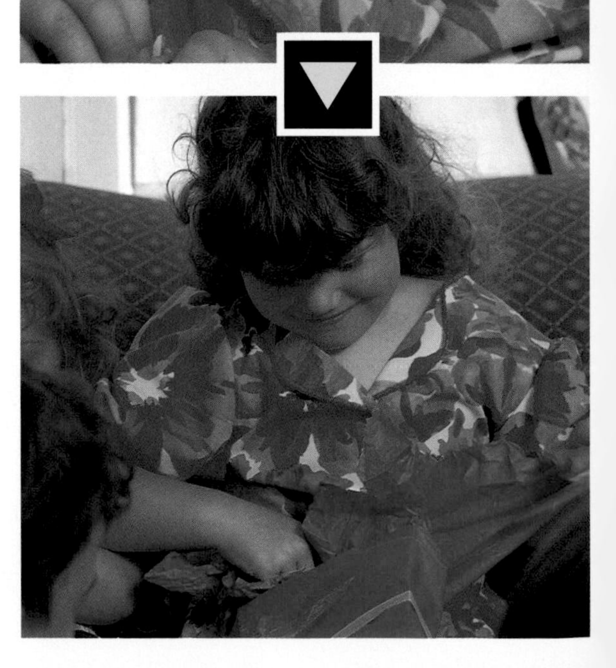

SOUND ADVICE

WHAT A RACKET!

With only your camcorder's built-in mic to rely on, the sound-track to your party is bound to be a somewhat chaotic affair, with lots of screams, laughter, shouts and gurgles. One thing you can remember is to keep quiet yourself during recording. When taping the children dancing, you could try standing near the hi-fi speakers, so that the music drowns out yells.

▶ Most of the children will be excited about being filmed and will pose happily with the Birthday Girl. The unwrapping of a present will distract their attention while you zoom in for some close-ups.

of the food and table settings. (The party organizers will certainly thank you later for this record of their achievements.) Before the guests arrive, be ready in a pre-determined position in the hallway to film them greeting the Birthday Boy or Girl. Record each of them as they move to the party room.

THE FUN BEGINS

With all the guests arrived (try to get close-ups of each one) let everyone work off surplus energy with a few party games, such as Statues, Musical Bumps or Musical Chairs. For Musical Chairs (or Cushions), start with your high, wide shot. The ideal position is then at either end of the line of chairs. When there are only a few chairs left, move 90 degrees to focus the children on one side. Move positions while the music is off and signal the tape operator when you are ready again.

A few cutaway shots of adult reactions to the children's antics may be especially amusing here.

As well as shooting the whole room from your vantage point on the chair or elsewhere, try getting right down to the kids' level. This low angle is very useful for good shots, if a little hard on the knees. A good way of keeping your recording activities concealed is to set your camcorder

◀ **The high points of your video will be the tea, the singing of 'Happy Birthday', and the cutting of the cake. Try to get a good close-up of each guest in turn.**

▶ **This high angle wide shot takes in the whole scene. Extra lights were set up to counteract natural light from the windows and ensure faces were properly lit.**

SHOOTING THE KIDS

Children are unpredictable: just when you think they are bound to do something entertaining, they do absolutely nothing at all. If a shot does not live up to your hopes rewind and re-cue. Or consider handing over the camcorder to the kids for a 'Funny Faces' sequence.

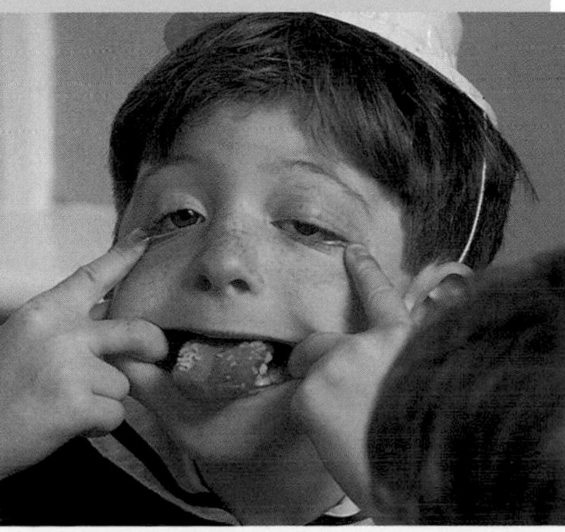

on wide angle and simply stand holding it low down at your side and let the action unfold in front of the lens. This technique of 'shooting blind' can be practised in advance; alternatively you can rotate the viewfinder to the upright position to check your framing.

TIME FOR TEA

The games over, it's time for the main event of the party – tea and birthday cake. Frame an establishing shot of the carefully laid tea table before the guests rush in and demolish what's on offer. Cut away to shots of plates of sandwiches being cleared or fairy cakes being bitten into. Be methodical here – go round the table in a clockwise direction taking separate shots of each of the children (all their parents will be glad to see the video later), starting and ending with the Birthday child. To provide visual cohesion to the sequence, try to make the shot sizes consistent – either using a series of close-ups or mid shots.

Now comes the important sequence – the cake. Start with a shot of Mum or Dad in another room putting the finishing touches to it. As the candles are lit, in close-up, the auto-iris will make continuous adjustment for correct exposure. Don't get too close – the heat could damage your camcorder.

TURN THE LIGHTS DOWN LOW

Once the candles are lit, return to your high angle, master shot position and give a visual signal for the cake to be brought in once you have started recording again. For this sequence, lighting levels can be reduced (lights turned off or curtains drawn), but beware of overdoing this – the auto-iris may simply compensate to equalize your

▶ **In the kitchen, get a good close-up of the cake being decorated. Take up your position in the dining room for a wide shot as the cake is brought in.**

changes. In order to record the whole of the Happy Birthday song, you will need to shoot without a break. So stay on a wide shot and zoom in towards the end for the blowing out of the candles. Then move swiftly to a closer position at the table-side for a side view of the cake-cutting. If anything goes badly wrong, just suggest that everyone has a go at doing it again. Most children love blowing out candles.

After tea, it's back to the main party room for a few rounds of a quiet game, such as Pass the Parcel. Seat the children in a circle. Once they are seated, get a wide shot of the circle. Then step inside and position yourself in the centre with the camcorder at ground level. The action will be too fast to follow completely, so simply concentrate on shots of two or three children at a time so that you really capture their expressions as the game progresses. From this position, you can gradually

▶ **Try a lower angle for the singing of 'Happy Birthday', holding for the blowing out of the candles. This gives a guest's eye view of the scene.**

GETTING ON DOWN

A sure way of bringing the viewer right into the action of the party when you play back your video on your television is to bend down and shoot the children at their head level – or even lower. This is particularly good for exciting close-ups of young children. If you want to catch them 'unawares', try recording them from a 45-degree angle while they are playing with a friend or looking at a toy.

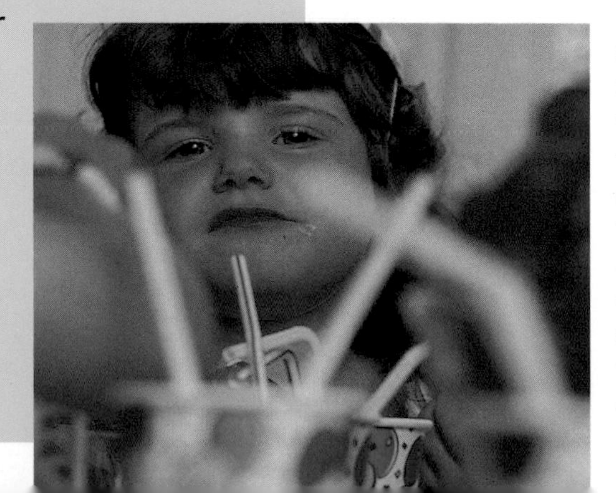

SHOWTIME

A children's entertainer will make your child's party that bit more memorable. The guests can let their tea go down enjoying a few conjuring tricks. Remember, however, to tell the entertainer in advance that you plan to video part of his act. A wide shot taken from behind the audience, who should be seated in a half circle with backs to the main source of light in the room will provide a good overall view of the action. You can then zoom in for shots of the key moments of stage 'magic'. Move to the side to obtain close-ups of enthralled faces.

rotate, until all have been included in the sequence.

For variation, you could try brief shots of torn wrapping or shots of the parcel, dwindling in size as it is unwrapped. Caught up in the game, the children should appear totally unselfconscious and unaware of the camera. It is an ideal opportunity to get a good close-up of each guest.

In the bag

By the time the games and tricks are finished, there may be time for a last dance before parents arrive. By now the party will be winding down, and to ensure that all the guests leave happy, a party bag each is essential.

Rather than just handing them out as guests leave, you could arrange for the correct number of bags to be hidden in a large tub full of bran or sand, turning the whole exercise into a Lucky Dip. You should be able to get some good shots of the children clustering round the tub for a good rummage and examining their bag's contents before they are ushered out by their parents.

Just as the party must come to an end, so must your video. If you are

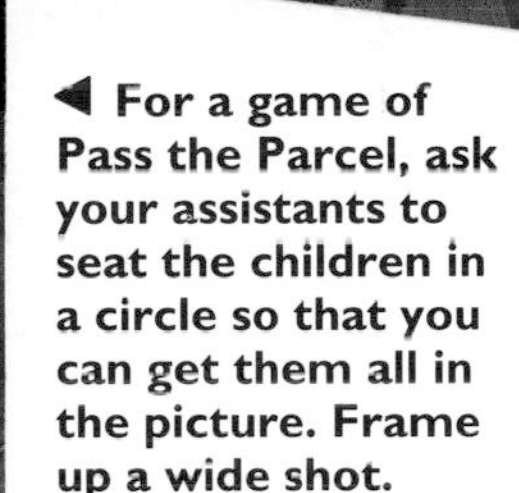

◀ **For a game of Pass the Parcel, ask your assistants to seat the children in a circle so that you can get them all in the picture. Frame up a wide shot.**

▶ **When the music stops, zoom in for a medium close-up to catch yet another layer of paper being ripped off. You could also try getting a close-up of fumbling hands. End the sequence with a shot of the unwrapped present.**

◀ A Lucky Dip tub full of party bags provides a focal point for action and a fun ending to the party with plenty of close-up opportunities.

▼ For some 'Goodbye' shots, you could try positioning yourself outside to get the Birthday Girl's reactions.

stuck for a suitable ending for your party video, you could try stepping outside your home and shooting the Birthday Boy or Girl waving goodbye to his or her friends from the front door of the house.

Goodbyee!

Start with a mid shot of your child waving at the door and then pull back slowly to fade out on a long shot of the scene. Alternatively, conclude with close-up shots of the gifts received with the child providing a live commentary, describing all the presents and who they came from.

By this time your home will probably look as if a bomb has hit it (and a shot of the debris could provide a visual commentary on the party's success) but, more important, you will have a video to treasure for years to come. □

BIRTHDAY BLUES

Plan your camera positions carefully beforehand to get the best possible shots

• **In the dark** – Don't shoot the kids against a window – your camcorder's automatic iris won't be able to cope.

• **Heads down** – Rewind and re-cue the tape at once if suddenly faced with backs of heads. There's nothing more boring for a viewer to watch. Don't shout orders at the kids: move your position or wait calmly until they look up again.

TRIPODS

A TRIPOD IS ARGUABLY THE MOST USEFUL ACCESSORY A CAMCORDER USER CAN BUY. NOT ONLY WILL IT GIVE YOU SMOOTH, SHAKE-FREE SEQUENCES EVERY TIME, IT GIVES YOU THE CHANCE TO GET INTO THE PICTURE, TOO!

A tripod is a three-legged stand designed to support a camera or camcorder, holding it steady so you can produce perfect shots every time. Although most of today's camcorders are designed to be hand held, and some are even fitted with anti-shake devices, there is no substitute for a solid base for your machine.

With a tripod you can achieve wobble-free pans and tilts, smooth zooms and record all kinds of events from a perfectly static standpoint. But more than simply improving the videos you already make, a tripod can lead you into the realms of creative filming, with macro, timelapse, trick effects and animation becoming real possibilities.

▲ **Pans, tilts and zooms take on a professional quality with the steadying effect of a tripod.**

▼ **Some models offer a special centre pole attachment – useful for shooting stills of documents and photographs, and invaluable for many kinds of macro recording.**

WEIGHTY MATTERS

Walk into any camcorder store and you will be faced with an enormous array of tripods, all looking extremely similar. Yet a quick glance at the prices will soon prove that there is quite a difference between them. To start with, you need to check whether the tripod was designed to accomodate a camcorder, or whether it has a camera mounting only. The camera mounting plate comprises a single screw, whereas the camcorder plate has both a screw and a retractable pin to keep it firmly in position.

Once you know that the tripod is compatible, two basic considerations will govern your choice: expense and

PHOTOGRAPHS BY RAY DUNS

ANATOMY OF A TRIPOD

Tripods have come a long way since the early days of the photographer's wood and brass supports and today's lightweight plastic and aluminium models are specially constructed to give the camcorder user a maximum number of facilities. Features vary from model to model – look out for the following. . .

THE CENTRE COLUMN For extra height and precise positioning, many tripods have an extendable centre column. This is operated by means of a crank handle which winds the head upwards then locks into position to prevent any further movement.

THE PAN HANDLE This enables you to pan and tilt the head as required. Most handles take the form of a single lever which twists to lock the head in position, but some camcorder tripods now have looped handles which give you a good grip.

THE HEAD AND MOUNTING The head of a tripod provides the movement, and all heads are designed to enable you to pivot the camcorder up, down and from side to side as smoothly as possible. In most cases it is the quality of the head which determines the price, and with top-of-the-range tripods the heads can be purchased separately from the legs.

The mounting plates have screw and pin fittings, and many can be instantly detatched with a flick of the quick release lever. This is a handy device as it means you can switch the camcorder from a mounted to a hand-held position in a matter of seconds.

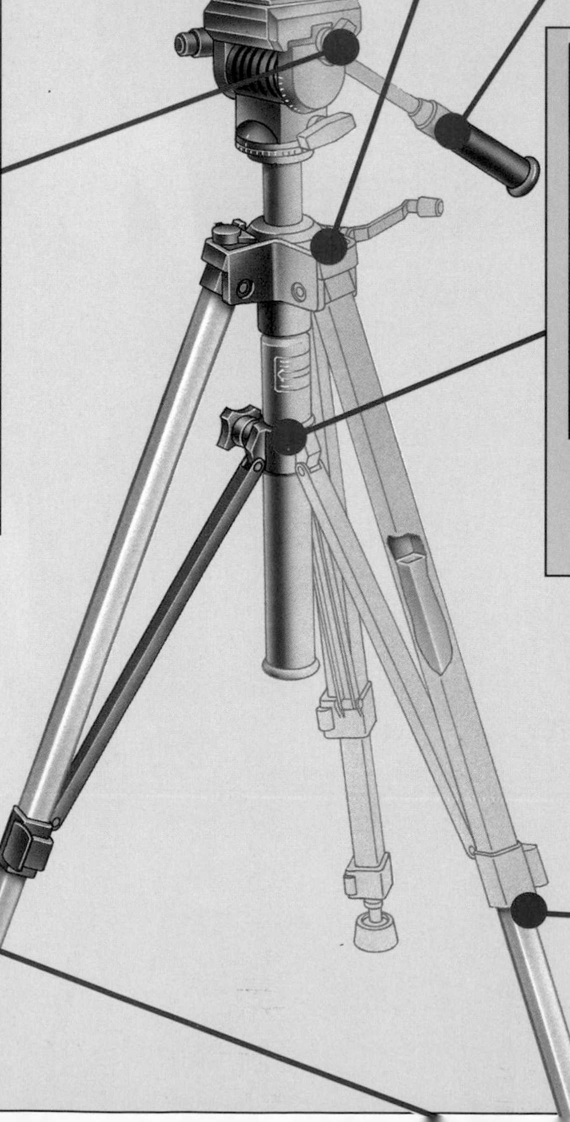

THE LOCKING NUT Legs need to be locked into position if the tripod is to be rock steady. This is often done with a nut which locks the struts to the centre column.

THE FEET Angled or ball-jointed for stability, tripod feet have non-slip rubber cushions with extendable metal spikes for extra grip on rough ground.

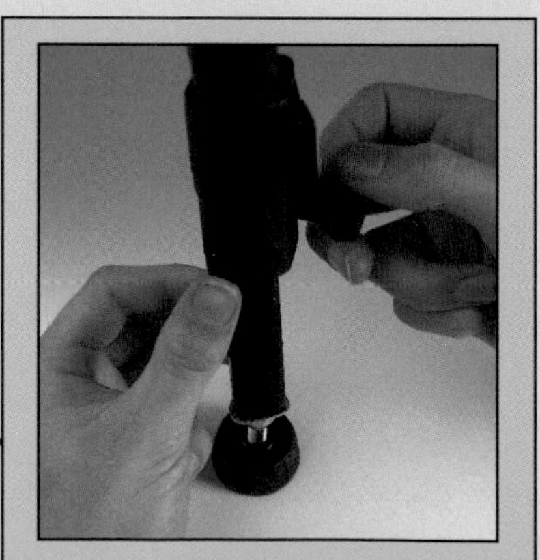

THE LEGS All tripod legs are extendable and are designed with lockable flip catches to enable you to adjust the tripod to the

ILLUSTRATION MARK FRANKLIN

weight. The heavier the tripod the steadier it will be, so it just depends on how much you are prepared to carry (the lightest camcorder models weigh around 1000g).

The basic materials are anodised aluminium alloys or rigid plastic – and most have telescopic, flip-lockable legs. Some have an extendable central column which can be handy, but this must be of a high quality if it is not going to wobble.

HEAD LINES

What does vary from tripod to tripod is the movable section at the top known as the head. It is the head which enables you to pan and tilt the camcorder, and therefore it should be as jerk-free as possible. There are three basic kinds of head. The simplest, and usually the cheapest, is the straightforward friction head. At the other end of the range is the fluid head, in which a liquid (graphite or silicon based) lubricates a series of ball bearings. These heads are often assembled by hand, and although the quality is excellent, they can be way beyond the reach of the amateur pocket. In the middle of the range are the fluid-effect heads, in which grease takes the place of the lubricant.

THE DOWN SIDE

There is no denying that tripods can be cumbersome pieces of equipment. Bags and clip-on

CAMERA SENSE

USING YOUR TRIPOD

- When you set up your tripod, lock the legs firmly and arrange them in such a way that you can stand comfortably in between.
- Ideally, the viewfinder of the camcorder should be positioned between eye and chest level otherwise you will find it uncomfortable to film.
- Lock the pan handle immediately after you have finished recording in order to prevent the camcorder toppling forward (or worse, backwards, as the lens could point directly at the sun, causing irreparable damage to its delicate sensory mechanism).
- For high shots or low shots, find something solid to stand or sit on – you might be in the same position for a long time.

LOW SHOTS

HIGH SHOTS

REGULAR SHOTS

straps do help to ease the burden if you are out videoing for the day, however they are still awkward accessories to carry around.

And of course there are times when you won't want (or be able to) set up your tripod – when wandering along the seafront or in an amusement park for example. But try shooting a car rally where you need perfect pans, or a picnic (where you feature in the video along with the rest of the family) and you will find that the the tripod is invaluable – truly the camcorder user's best friend. □

CHECK IT OUT

- Weight and portability – don't buy a tripod too light for your particular camcorder or too heavy for you to carry.
- Compatibility – some older tripods designed for cameras may have a mounting screw only and no stabilising pin. Likewise beware of old camera tripods discovered in lofts or in the back of cupboards– they may appear to fit, but many have screws which are too long and if forced, could damage the camcorder's internal mechanism.
- Height – the size of the tripod will determine the height it will reach. Make sure the tripod you choose is tall enough for your needs.
- Low shots – there are several occasions when you will need to film close to the ground. Splay the legs to check how low the tripod will go.

Hold the shot for a few seconds before the moving subject comes into view then start the pan slowly.

Let subject reach the centre of your viewfinder, then follow its progress at the same speed, using the pan handle to swivel the tripod head.

Hold subject for 5-10 seconds, then let it drift out of shot, slowing but not ceasing the pan.

THE PERFECT PAN

The horizontal pan is one of the most basic, and if used appropriately, one of the most effective camera moves. Taken from the word panoramic, it is simply a smooth, slow swivel from a static point which enables the viewer to take in a scene in much the same way as a slow turn of the head. The pan is a perfect device for giving a full view of a long building, a city or landscape, or a moving object.

Smooth, hand-held pans are notoriously difficult to achieve, whereas pans made with the aid of a tripod are simplicity itself. The trick is to take it slowly – often slower than you think you need to go. Practice should give you perfect results.

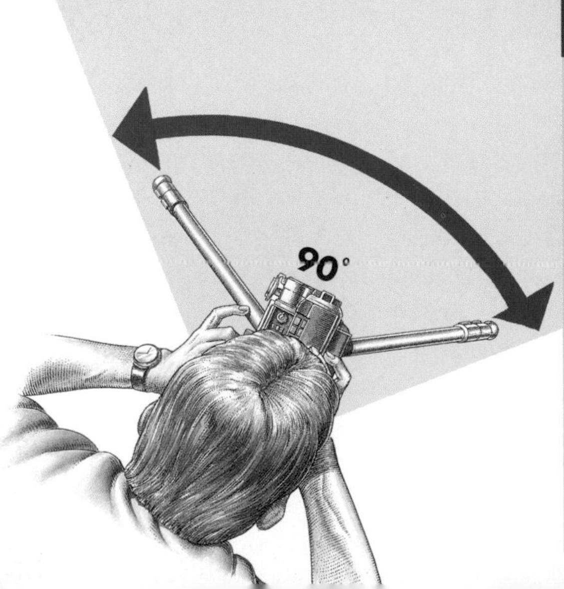

ILLUSTRATION BILL LE FEVER

CAMCORDER CARE

WARNING! RELIABLE AS IT MAY BE, A CAMCORDER NEEDS ATTENTION. SAVE YOURSELF MONEY, TIME AND POSSIBLE FRUSTRATION BY FOLLOWING A FEW SIMPLE RULES OF CARE THAT WILL HELP YOU LOOK AFTER YOUR EQUIPMENT.

A poor workman always blames his tools. How often have we had to resist actual physical violence when we hear this smug saying, as we play back quavering cassette recordings of a much-anticipated radio concert, or a flickering videotape of a favourite film, and curse the particular machine that 'let us down'.

Part of our rage, of course, stems from the fact that the saying may well be true. Without proper care and maintenance, electrical goods will go into gradual decline and fail to function to the optimum. And a camcorder is no exception to the rule. Negligence and maltreatment account for a large proportion of glitches and breakdowns.

But there are still more urgent reasons, apart from loss of recording quality, why you should take good care of your camcorder. Most damage is caused by knocks to the body-shell; by liquid, smoke or sand getting into the workings, or by wrenched lenses: and none of these are covered by the camcorder guarantee, nor by any extended guarantee you may buy at the time of purchase. Moreover, repairs to camcorders will often run into three figures for the specialist attention they demand.

Which parts of my camcorder will suffer from the most wear ?

Your camcorder's heads and motors. These should last for at least 1500 running hours.

A

Three basic precautions will keep your camcorder in good condition. First, do not allow moisture to enter the body-shell. If you are taping in the rain, make sure that the camcorder is covered with a 'rain jacket' (available on order from all major retailers). But the prime cause of moisture getting inside the works is condensation. This forms when the camcorder is moved from a cold to a warm place, and may damage both the tape and the circuitry. If filming outside in winter, remove the video cassette and place the camera in a plastic bag prior to taking it into a warm room. Once in the warm, wait until the camcorder no longer feels cold to the touch before removing it from the bag.

Dust also creates problems if it gets inside your equipment, so keep your camcorder and accessories in their cases when not in use.

ILLUSTRATION BY DAVID ASHFORD

▲ **A camcorder enclosed in a rainjacket – an important accessory if you plan to do a lot of outdoor shooting.**

Secondly, do not subject the camcorder to extremes of heat or cold. Don't leave it lying next to a camp fire, or in the snow, without any protection! But above all, don't let the sun shine directly in the lens. This can cause curious tadpole shapes to appear on your film: the plastic surround to the viewfinder screen is burned by light focused through the viewfinder.

Safe keeping

Finally, do not subject the camcorder to hard knocks! Again, always put it away in its case after use. Inevitably, your camcorder will be most at risk as it is transported from place to place. So carry it in a well-padded bag. In the car, the camcorder should be wedged into a safe place on a seat, not put on the floor or in the boot where it may suffer vibration.

These measures should protect the camcorder from physical injury. Equally vital is the protection of your video quality. A common cause of poor picture quality is failure to keep the camcorder clean. There is nothing worse than returning from a tropical holiday to find that it looks on video like two weeks spent in a Siberian snowstorm. The snowstorm effect arises from dirty video heads, as do flickering and splotchy pictures. Heads can be cleaned with a cleaning tape. In addition, make sure your video tapes are in good condition; worn or faulty tapes are often the culprits when it comes to spoiled recordings.

Lens cleaning

The lens is another feature which should be kept clean. All camcorder lenses attract dust, no matter what precautions you take. Dust can be removed using a blower-brush, which should be worked outwards from the centre of the lens to the edges. Fingers, of course, should be kept clear of the lens at all times. If you do smear it, apply lens-cleaning fluid with lens tissues.

A camcorder doesn't have to be handled like a Ming vase, but it does deserve respect. On the whole, it is a remarkably reliable piece of equipment. Providing you handle and store it carefully and keep it clean, it is estimated that, like most electrical equipment, it will give you at least seven years of good service. □

▼ **An array of camcorder and VCR head cleaners. The camcorder head cleaners should be used after 30 hours of playback/recording or after outdoor shooting to maintain good performance.**

Zoom in!

PHOTOGRAPHS BY LYNDON PARKER

ZOOMING IS A TECHNIQUE THAT CAN MAKE OR BREAK YOUR RECORDINGS. MASTER THE ART AND OPEN UP A WHOLE NEW RANGE OF VIDEO POSSIBILITIES

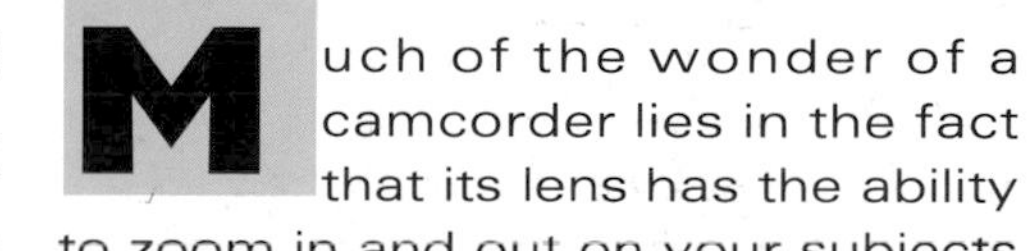

Much of the wonder of a camcorder lies in the fact that its lens has the ability to zoom in and out on your subjects at the simple press of a button. This is one of its greatest advantages, and also one of its greatest drawbacks. The tendency to zoom in quickly from a well-framed wide shot to a close-up, then straight back out again, is a common one. It is usually a feature of most people's first home videos and has the result of making the viewer feel slightly seasick when they watch it on TV!

The zoom lens is a marvellous feature and can add a great deal to the creative quality of your recordings – if it is used properly. But in general, a little zooming tends to

LEARN HOW TO ▶▶ *Use the zoom between shots to frame your scenes* ▶ *Zoom and tilt for the right shot* ▶ *Pre-plan your zoom shots on Standby*

◀ **HAND HELD ON WIDE ANGLE** **This shot of a family picnic illustrates one of the fundamental problems of hand-held zooming. The shot begins well. It is correctly framed and on the widest angle there is minimal camera shake.**

◀ **HAND HELD ON TELEPHOTO** **As you zoom in to telephoto, the image becomes increasingly unstable.**

▶ **TELEPHOTO USING TRIPOD** **The same shot, taken with a tripod. The camera shake common to hand-held telephoto images – caused by unsteady hands – has disappeared.**

go a long way in most family videos, and knowing exactly when to use this facility and when to hold back is all part of the art.

What is a zoom lens?

In order to make full creative use of the zoom, it is important to understand exactly what it is. The term 'zoom lens' (which derives from the company name of one of the early manufacturers) is, in fact, a misnomer which is largely responsible for misleading camcorder users about how the lens should be employed.

The lens is in fact a 'variable focal length' lens. (The focal length is the distance between the optical centre of the lens and the point at which the lens is focused.) The zoom lens was originally developed for feature film-makers in the 1940s to enable them to adjust shot sizes without having to move the camera itself or change the lens. In the old Hollywood studios, time was money and any device that could speed up production was worth developing. It was only with documentary and news material that film-makers found the justification for employing the zoom function during a shot.

How does it work?

A camcorder's zoom lens is usually described by manufacturers in terms of its zoom range – that is the ratio of the shortest and longest focal length of the lens. These lengths, or the two extremes of the range, may be marked on the barrel. A typical camcorder zoom with a range of 7mm to 56mm will be called an 8:1 lens or 8x; others may have ranges of 10x or even more.

The focal length is important because it determines the magnifying power of the lens. The longer the focal length, the greater the magnification of the image. The shorter the focal length, the less the magnification.

CAMERA SENSE

ROOM FOR MANOEUVRE

An understanding of exactly what the camcorder lens can show you in your recordings will help considerably when planning and shooting your videos. Try this exercise in familiar surroundings:

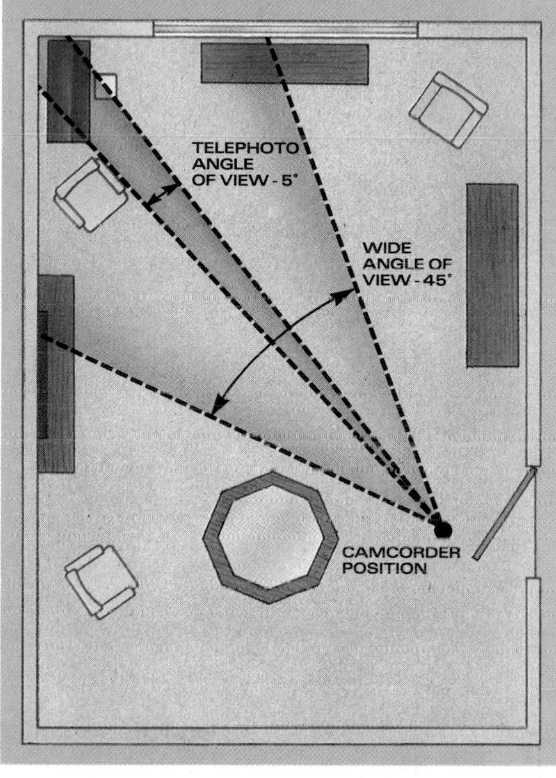

DIAGRAM BY MARK FRANKLIN

- **Set your camcorder up in the corner of the room.**
- **Take a look at the room's actual size.**
- **Now look to see how much is included with the lens set to wide.**
- **Zoom in to maximum telephoto and note the image size.**
- **Try to remember these extremes when planning to shoot in different circumstances.**

The focal length is also significant in terms of the angle of view that the lens shows. A typical camcorder zoom lens will show an angle of around 45 degrees at its shortest focal length (the 'wide angle' position) and around 5 degrees at the longest focal length (the 'telephoto' end).

Zoom settings

Since the zoom lens incorporates both a 'wide angle' and a 'telephoto' position, it is capable of producing a range of phenomena familiar to most amateur photographers. At the widest setting, the view will be similar to that perceived by the human eye, with foreground and background subjects seen as being distant from one another and in natural perspective. At the telephoto end, the image appears flattened and space between the objects in your viewfinder compressed.

The comparative narrowness even of the 'wide angle' position means you cannot always get as much into a shot as you had hoped – particularly when shooting indoors. You will often have to 'pan' the camera across a room to show its full size. On the other hand, the convenience of the zoom lens' telephoto position can tempt you into shooting tight, close shots from some distance away. Without a tripod this can cause wobbly shots.

Speed of movement

On all camcorders the zoom function is motorized and operated via a simple rocker switch, using the first and second fingers of your hand. The zoom speed is governed by this motor and is usually set to operate at around 8 to 10 seconds to cover the entire zoom range. Some camcorders also have a faster set speed, and the most sophisticated machines enable you to vary the zoom speed according to the pressure exerted on the switch – in much the same way that they do on professional video cameras.

Although the motorized zoom can ensure smooth zoom shots, the set speed can sometimes be a problem, particularly when covering action subjects. For instance, suppose you are covering a motor race or go-kart event and set up a telephoto shot of the vehicles racing towards you. When you start zooming out to keep them in shot as they come closer, you may find that they are moving faster than your zoom speed will allow and they disappear from shot before they pass you. Similar problems can occur with less predictable subjects, such as a dog playing in the park.

So the first thing to understand about your zoom lens is that it is like

Q ***I have been trying to video my young kittens' antics but every time I zoom in for a close-up they have moved out of shot too soon.***

The speed of your zoom motor is just not quick enough for this situation. Use manual zoom so that you can change your shot size quickly to suit their position. **A**

having a whole set of different lenses of varying focal lengths in a single compact unit. Each time you prepare a shot, you simply adjust the lens to produce the best possible composition from that camera position. When you finish that shot, stop recording and move to another camera position – or simply adjust the focal length of the lens to compose a new shot and start recording again. In this way most of your 'zooming' will actually take place between and not during shots – improving the look of your video.

IMPROVE YOUR SKILLS

Play the Game

In order to achieve the perfect zoom, it is often necessary to combine it with another movement such as a pan or a tilt. This ensures proper framing from start to finish.

▲ STATIC ZOOM
If you start with a close-up then simply zoom out, making no additional camera movement, your shot will end up badly framed.
◀ The result is too much headroom and little indication of what is going on.

▶ THE TILT
A smooth, slow tilt downwards while zooming out will reveal the game that the man and his son are playing.

◀ THE PAN
Camera movements can make a difference when you zoom in, too. This time you will need to pan to the right as the focal length increases.

▲ TELEPHOTO SHOT
Finish with a close-up of the boy's face. It is a good idea to rehearse the move in advance to ensure successful framing.

CAMERA SENSE

DIGITAL ZOOM

Some camcorders are equipped with a digital zoom facility. This is capable of enlarging the centre part of the screen image and has the effect of doubling (or more) the maximum focal length of the lens. However, there will be some deterioration in picture quality.

You can test the adage 'Zoom between shots, not during them' with a simple exercise in your living room. Set the camcorder up in one corner, preferably on a tripod, and record a 10-second static wide shot of the room. Then, without moving position, compose a second shot using a longer focal length of, say, the sofa, and record 5 seconds.

Now compose a third shot (still without moving the camera position), again using a still longer focal length to provide a close shot of a single chair and record 3 seconds. Continue in the same manner, choosing shots of three other objects (a vase of flowers; a picture on the wall; a bookshelf) and record 3 seconds of each. Finish the series of shots with one framed, say, looking out of the window and hold this for 5 seconds.

THE SECOND VERSION

Next, start a second recording, again commencing with the wide view of the room held for 10 seconds. This time, without stopping recording, continue shooting by zooming in and out on each of the same subjects you covered, trying to duplicate the composition you used and holding each of the subjects for the same amount of time. Finish the shot with the view from the window.

ARCHITECTURAL IMAGES

Watch any TV documentary on the subject of architecture and you will see examples of zoom shots used to brilliant effect. When shooting your own videos, be selective: some subjects work better than others. Try to choose subjects which look good at both ends of the focal spectrum. This rose window is a good example as its circular form works both in telephoto and in wide angle.

Play back the result on your TV and decide for yourself which version you prefer – the sequence of separate shots or the continuous recording with the camcorder continually zooming and moving to show what is in the room. It is highly unlikely that you will find this second version of the 'portrait' of your own living room as easy to watch as the first. You will thereby have proven the value of thinking of the zoom lens as a variable focal length lens.

You will also have noticed several other points from this experiment. Most importantly, that it is possible to shoot a number of different shots from a single camera position simply by using the range of the zoom lens' focal lengths. Although you should never feel that you should not move your camera position – and the portrait of your room would have undoubtedly looked better if you had changed angle for each shot in the first version – there will be times when your ability to move with your camcorder is restricted. But by varying the focal lengths of the zoom lens you can still create a good sequence, made up of shots of different sizes.

Finally, if you were not able to shoot your video using a tripod (or other camera support), you will have plenty of evidence of the tricky problem of camera shake when using the camcorder in anything other than the smallest focal length and the widest angle of view. None of this, however, should deter you from including zoom shots in your home videos – particularly when covering action events or to make stationary subjects more visually interesting.

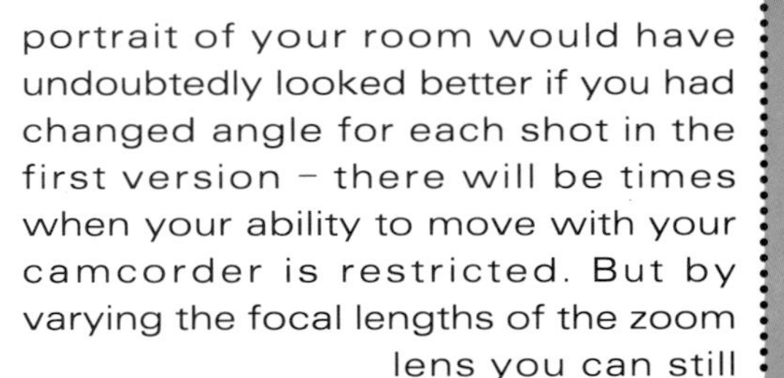

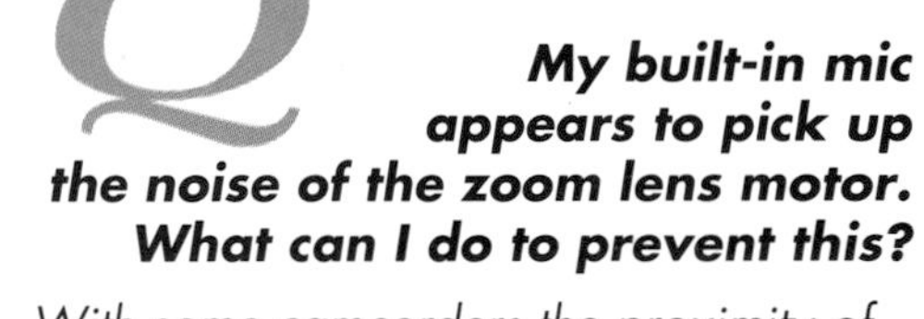

Q ***My built-in mic appears to pick up the noise of the zoom lens motor. What can I do to prevent this?***

A *With some camcorders the proximity of the zoom motor to the built-in mic makes this a real problem. A solution may be to fit an auxiliary directional mic to the accessory shoe.*

CHECK YOUR ZOOMING
The problem of zooming out is that things either appear out of nowhere or get cut off! Check the full shot before you record.

• **Linen folds** – A nicely composed shot, until the zoom out caught the washing line to the side of the frame

• **Half the house** – A zoom out from the door, but it didn't go far enough!

• **Window worries** – Zooming out, the glass has confused the autofocus

PHOTOGRAPH BY RONALD SHERIDAN / THE ANCIENT ART AND ARCHITECTURE COLLECTION

The editor's eye

ILLUSTRATION BY NICK PAYNE/THE GARDEN STUDIO

SKILFUL EDITING CAN TURN ALMOST ANY VIDEO INTO A VISUAL TREAT. BUT BEFORE YOU CONNECT UP YOUR EQUIPMENT, TRY LOOKING AT YOUR TAPES WITH AN EDITOR'S EYE

Even seasoned professionals sometimes shoot material that does not work well or that contains mistakes or retakes. The craft of editing is, at its most basic, the process of choosing and keeping the best of what has been shot. All that matters is that buried amid the accidents, technical hitches and scenes that did not live up to your expectations lie the key shots that you wanted to capture all along. The only way that you will be able to identify these nuggets, however, is to watch your tape over and over again, until you know it warts and all.

Suppose you have made a video of your family's day out at the seaside. When you view it later, you will probably have mixed emotions about what you have recorded. That section half way through where the camcorder was accidentally switched on, resulting in a sequence of feet trudging along the sand seemed funny the first time, but boring the second and infuriating the third. However, you could well decide that this accidental shot is just in the wrong place at present and might make an interestingly 'spontaneous' opening to your video.

WORKING ON PAPER

As you view, it is a very good idea to keep track of what you are watching by noting down the effectiveness of each shot and the tape counter number at its beginning and its end. You will thus gradually build up a list on paper of everything on the tape, its precise location and usefulness. Professionals call these notes an 'edit log'. This is a very

▲ Take plenty of time to review all the tape you have shot on your television, carefully noting down what you see. This is the vital first step in planning to edit any video.

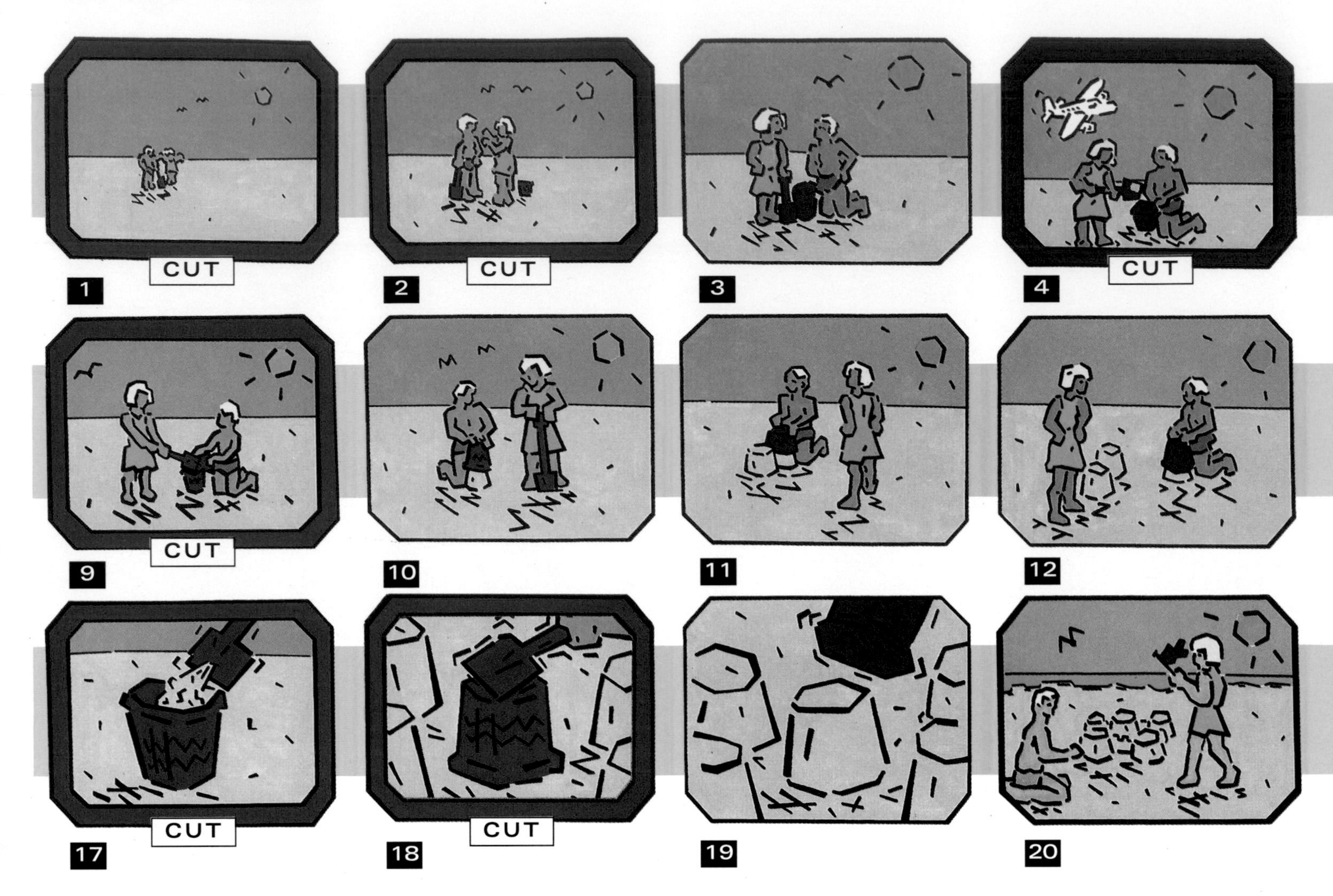

▲ COMPRESSING A SEQUENCE
This 'episode' from a day at the seaside shows how to plan and cut out technical errors (frames with red borders) to improve the result. The children are too far away in shots 1 and 2; shot 3 makes a good starting point. In 4 and 5 an aircraft shatters the mood, so cut from 3 to the close-up of 6, returning to the long shot 10 and holding through 11, 12, 13, and 14. This shot is held to capture the whole of a song the children sing as they build. (Continued on opposite page.)

'time-consuming business probably best approached in short bursts of an hour or so at a time to prevent serious headaches!

From your notes, you will see that your tape can be divided into several episodes – in effect, your video may well be made up of five or six little films. Once you have identified these episodes, your job as editor is to make sure that each one has a beginning, middle and end.

Interest value

In general, each episode should begin with a long shot that sets the scene. You can then briefly introduce the personalities who will figure in the action, placing them in context. Judge the length of each shot in terms of its interest value. Ensure that every shot serves a definite purpose.

You may well feel that the action takes much too long to develop at certain points. Look through your notes to see if there is another shot you could insert (perhaps a close-up of someone's reaction), which would allow you to return to your main action at a more interesting point. In this way you will avoid a disconcerting, jarring effect called a 'jump cut'. Similarly, avoid planning a cut that will make it appear that a person or thing has jumped from one side of the screen to another, or a cut that suddenly places the same person against a completely different background.

If your video contains a number of zooms or pans you may be faced with some difficult decisions: it is a bad error to plan a cut in the middle of a pan or zoom, so assess how well each pan or zoom works in context before choosing whether to leave it in or take it out entirely. Use 'tricksy' shots such as pans, zooms or unusual angles sparingly – too many can be tiring to watch.

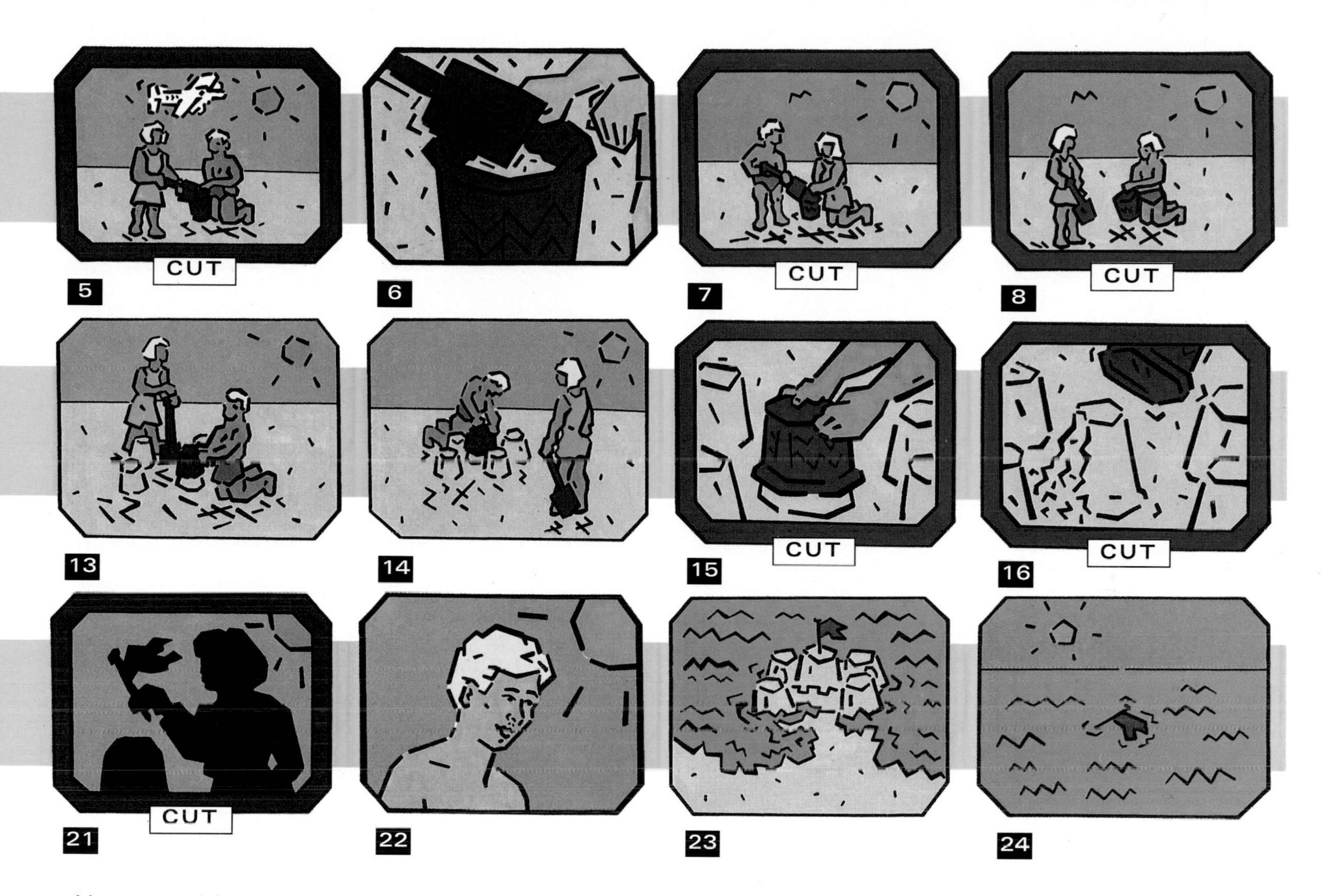

You may wish to compress a sequence of actions, for example, the children looking for sea-shells, throwing stones at the waves, then taking their shoes off and paddling. The simplest way to cut this down would be to select just one of these actions, carrying it through from beginning to end.

THE RIGHT QUESTIONS

As you work through your tape ask yourself a series of questions. Will the viewer understand where the action is about to take place? Does the action take too long to develop? Does each sequence have an effective resolution that will allow a smooth transition to the next one?

When you have reviewed each sequence of your video, spotted serious technical errors and planned ways to 'tighten up' the rest, you will have begun to realize the possibilities that editing opens up to the videomaker. For the secret of making an effective video lies as much, if not more, in skilful 'post-production' editing as in careful pre-planning and shooting.

Cutaway shots are short images that cut away for a few seconds from the main action to show people's reactions (known as a 'reaction shot') or subjects relevant to the context of the main action. These shots can be worked out in-camera or planned in at the editing stage.

An extremely common use of the cutaway shot is to compress time in a sequence. For example, you might wish to edit a lengthy sequence of two people walking towards some rock pools. After a few seconds of this particular image, you could plan a brief shot of gulls wheeling in the sky, returning to your subjects only when they reach the rock pools. The cutaway will have added interest and also conveyed that the walk to the rock

▲ A close-up of a sandcastle being created (15, 16, 17, 18) does not work but the 'retake' (19) does. The tide is coming in (20); the children do not notice. The lighting error of 21, which occurred as the girl placed a flag on the centre sandcastle can be planned out by cutting straight to the boy's reaction (22). What he is looking at is immediately clear as the sea washes over the castle (23 and 24). Turn the page to see how the final edit will look.

THE 'FINAL EDIT'

▲ **With the technical errors edited out, the sandcastle episode will have a tighter structure with a clear beginning middle and climax when the castle is suddenly swamped by the incoming tide.**

SOUND ADVICE

LISTEN CAREFULLY . . .

As you run through your tape making your editing notes, listen carefully to the video's sound-track as you watch. There may be times when a particular shot does not work as well as you might like visually, but it may contain an entertaining sound or vital snatch of dialogue. Unless you are planning to get rid of your 'live' sound altogether and re-create a new musical accompaniment to your video, you will have to retain all the sequences that feature important sound.

pools took your subjects a certain amount of time. Useful as they are, cutaway and reaction shots should not be overused – too frequent a use can distract from the main action and be irritating for the viewer.

SHOOTING TO EDIT

Making a video of a day out knowing that you are going to edit your tapes allows you to experiment. For example, you can record the same subject from several different positions. There is no need to record over mistakes – in any case, they may look better than you thought when played back on TV. The more you shoot, the more choices you leave yourself when editing.

Editing tapes by selecting what you want and copying it on to another tape means that you do not even have to worry about the order you shoot sequences in. You can leave all considerations of shaping your material – and really creating your video – for later. □

On Safari!

LIONS, TIGERS, YOUR FAMILY AND THE GREAT OUTDOORS – ALL THIS IS THE STUFF OF A BRILLIANT HOME VIDEO. SO CHARGE UP THE BATTERIES, ROUND UP THE FAMILY, AND GO BIG-GAME HUNTING WITH YOUR CAMCORDER

There are few better places to make a memorable family video than a safari theme park with all its exotic animals to amuse and delight your children. But if you want your video to give as much enjoyment in ten years' time as it does when played back on your TV today, you will need to do a bit more than just point and shoot.

Every trip out with the family and every theme park will be different. Even if you know the park like the back of your hand, there are still the children, the other visitors and the animals to consider, in addition to the weather. With so many variables, where do you start?

Time and place

It will help considerably if you are familiar with the park. If not, try to borrow some leaflets from a friend who has already been, or ask the park to send you some information in advance. A quick read through should be enough for you to work out what you want to see and when.

Having an itinerary in your mind will help enormously. Some of the bigger parks have so much to see that it's impossible to do it all in a day, so you can weed out the things you are likely to find less interesting to view later on video. Try to imagine your trip as a short film in your head with a beginning, middle and end.

It is not always possible to get the necessary information in advance, so if you have to pick up leaflets at the gate, spend a few minutes studying them – particularly the feeding times as this is often when the animals are at their most lively.

Before you set off in the car, establish who's going to do the shooting; you cannot drive and record at the same time! (The front passenger seat is the ideal shooting position, giving you access to both the windscreen and the passenger window.) Hold the camcorder in both hands, with your elbows pressed in against your body. In order to obtain maximum steadiness, brace yourself firmly against the seat or the car door.

A sequence that shows the family's arrival makes a good beginning to the day's filming. Roadsigns, the park sign and a shot of the family car driving through the entry gate make good viewing. If you want to get a shot of a road direction sign from the car, make sure you are moving slowly enough, otherwise the sign will be an unreadable blur.

PLANNING CHECKLIST

- **Make sure that your camcorder has enough battery power to last the whole day. Charge them the day before**
- **Take an umbrella (or a camcorder rain jacket if you have one) so you can continue to shoot in the drizzle if you want**
- **Pack at least one spare tape**
- **Don't pack the camcorder in the boot – you may need it in the car on the way there**
- **Familiarize yourself with the layout of the park and the available activities in advance and work out a rough itinerary**

▲ Set the scene with a sequence of the car driving through the safari park gate. Keep the action flowing in the same direction, otherwise it will look as if the car is driving out again!

An opening title

You can always stop near the entrance gate to get a proper shot of the park sign that will serve as a 'title' for your video. If possible go ahead of your vehicle and record the car driving through the entrance gate and follow this with an establishing shot of the park.

The first thing you will probably want to see is the big game reserve,

INSERTS FOR EDITING

If you plan to edit your videos at a later stage, it's well worth picking up leaflets and brochures on your day out, as you can shoot any interesting logos, titles or still pictures and insert them into your video.

PHOTOGRAPHS BY STEVE LYNE/INSET TELEGRAPH COLOUR LIBRARY/SPECIAL THANKS TO WINDSOR SAFARI PARK

THROUGH THE WINDOW

To make the best recording from your car:

- **Remove all papers (and anything else, such as light clothing, that could cause reflections and confuse the autofocus while shooting through the windscreen) from the dashboard shelf.**
- **Keep the lens close to the glass. This will help eliminate reflections and enable the autofocus to fix upon the animals beyond the glass.**
- **Rain and drizzle on windows can spell disaster for the autofocus – set camcorder on Manual and focus on the animals outside.**

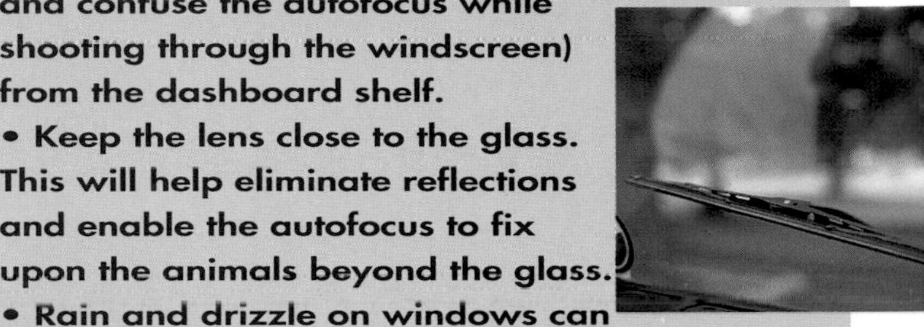

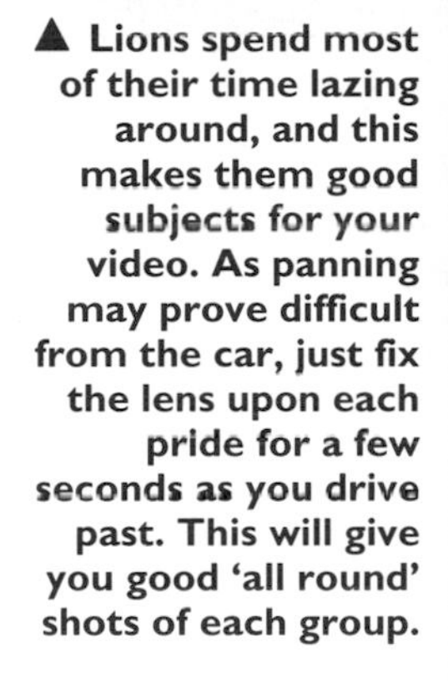

▲ Lions spend most of their time lazing around, and this makes them good subjects for your video. As panning may prove difficult from the car, just fix the lens upon each pride for a few seconds as you drive past. This will give you good 'all round' shots of each group.

full of dangerous animals. They are kept in fenced-off sections of the park, with electronic gates and cattle grids to stop them getting out. As you drive in you will see notices urging you not to stop, and to keep your windows and sun roof shut, and this you must do, as most of these beasts really would eat you alive.

▶ Reaction shots are important here – swivel round and record the children looking out of the car window. Then cut to an animal, a wolf or an exotic tiger, to give the feeling that you are driving past.

SHOOTING ON THE MOVE

Wildlife photographers spend years learning their trade and months obtaining the right shots. Weeks are then spent editing their footage to create all of the amazing nature programmes we see on TV. Where the lions, tigers, bears and rhinos in the safari park are concerned, you will have just seconds to get the right shot, through the windows of a moving vehicle. But whereas you can't expect to take award-winning sequences with your camcorder, a combination of skill and luck can provide you with impressive results.

This is where teamwork comes into its own. With your eye glued to

▲ **Follow a long shot of the family walking through the theme village with a close-up of the children's faces. Their gazes will lead you into the next shot.**

the lens, and sitting comfortably in the front seat, you will only be able to see what the camcorder sees. It is up to the rest of the family to yell 'tiger to the right' and 'zebra to the left' while you follow instructions.

Depending on the crowds, you can always drive around the reserve a few times, getting different shots with each circuit.

Sound problems

The biggest problem you are likely to encounter when shooting from the car is sound. If you manage to get that priceless shot of the tiger strolling across the road in front of you, when you play it back on the TV it will probably be accompanied by a running car engine and shrieks of delight from the whole family. If you don't want accompanying oohs and aahs as a sound-track to your video, there are two solutions. The first is to dub music over this section of the tape. The other, DIY option, is to quieten the family and play a music cassette in the car to drown out the

◀ **There is no need to miss out on the fun – hop on the ride in front of the children and record their reactions as the carousel whizzes round and round.**

SOUND ADVICE

WEATHER REPORT

If you are unlucky enough to get caught in a rainstorm, stop recording. Not only do most camcorders react badly to damp conditions, but any drops of rain that fall upon your built-in mic will create a loud crashing sound on your audio recording. Solve the problem by recording underneath an umbrella, or put your camcorder in a special rainproof case.

car engine. Alternatively you could simply accept the family's gleeful instructions as a natural commentary to the 'wild animal' film.

The fun of the fair

Most parks provide much more than just exotic and dangerous animals, and once you have driven round the reserve you will probably want to sample the amusements. This is where your family takes on the starring role. Remember, while you are standing still, getting your establishing shots, they will be wandering on ahead looking at all the fun the park has to offer. If they appear in the viewfinder at all you will see no more than their backs as they disappear into the distance. You will have to do a lot of running on ahead to get the right sequences.

Once you have positioned yourself well ahead to get a good establishing shot, just hold the camcorder still and let them walk slowly towards you. Hold the shot until the family have stopped to look at something of interest, then cut before swinging round to record whatever it was that caught their attention. Cutting back to a closer shot of the family looking in that same direction will complete a simple sequence.

In a spin

Fairground rides, always a source of delight for children, should provide you with some wonderful video recordings. If they all race off towards the carousel-style rides, stand back and shoot the whole machine – you will catch the kids on the way round. Alternatively, you could join in the fun yourself. Hop in the carriage in front, then turn round and shoot the kids in the seats behind you – you should get some great expressions.

Off the fairground rides and on to the animals again! Elephants are a perennial favourite and the opportunities here for good shots

▲ Slides and helter-skelters make great subjects for a tilt shot. Start shooting at the bottom and tilt slowly upwards, pausing for a couple of seconds at the top. Then move to a closer position and wait for the action (and fun!) to start.

▶ Positioned comfortably, carry on recording to capture the glee and excitement of a child as she spills out of the bottom of the ride. Keep the camera lens as wide as possible, moving nearer if necessary rather than zooming in, in order to avoid camera shake.

▲ **Frame an establishing shot to set the scene, then cut to a close-up. Follow with a mid-shot of the penguins' antics on the rocks.**

▶ **Slow-moving elephants make video recording easy – mix static shots with gentle pans for best effect.**

are tremendous. Most elephants are now kept in their own compound, in which case you will have to select your position carefully. Look for a position where you can get a high shot of the whole area to set the scene, then move down to the barrier. Take up a position some way ahead of the animals but well back from their path. You can record them lumbering towards you, then gently pan round with them as they pass.

CAMERA SENSE

BEHIND BARS

Most of the park's smaller animals will be kept in cages, and this can present problems for the videomaker. The metal bars, netting and wire mesh will mislead the autofocus of your camcorder into focusing on them – not the animals behind. If you cannot manage to shoot between the bars, take a few steps back and switch your lens to Manual focus, centring on the animal itself, until the mesh disappears out of focus.

Try shooting this from a low angle, kneeling on one leg to dramatize the size of the elephants. If you are really lucky, rides may be available for children: these make delightful sequences, as you can get some genuine interaction between the kids and the animals.

OUT TO LUNCH

Biscuits and birds next. Food is always a big part of any day out, and you can either choose to record a short sequence of 'lunch', whether it is a picnic or a takeaway burger, or go for simple shots of one member of the family eating an ice-cream. Ice-

ZOOM TIME

Animals in repose, such as these lemurs, make a great subject for a video zoom. Find some means of temporary support for your camcorder if you can, then do a trial run, moving from a wide shot to a tight telephoto position to ensure that you will get the shot you want. Adjust your position if necessary, then press Record and repeat the same process, going into a close-up of one of the animals.

cream shots have long been a favourite with photographers, and with a camcorder you can bring those images to life. Try to video the child unawares, so that they don't 'perform' for the camera.

Making a splash

Lunch over, you can move on to the smaller animals. Penguins are happy to perform for anyone, and if you catch them at feeding time you should get amusing sequences, but don't get too close, or you may end up with water on your lens. If you manage to catch feeding time, start with shots of birds waiting excitedly before cutting to the food being prepared by the keeper, to build up a sense of expectation. Then find a stable camera position so that you can concentrate on close views of food being eaten. But beware – your autofocus may have problems with the water and its reflections. Keep the birds in the centre of the frame or use manual focus to help prevent this.

Cutaways – brief shots of your family and the rest of the crowd reacting to the proceedings – become particularly important at this stage of

▶ On to another enclosure – by resting the camcorder on a fence or wall for steadiness, you can zoom in for good close shots.

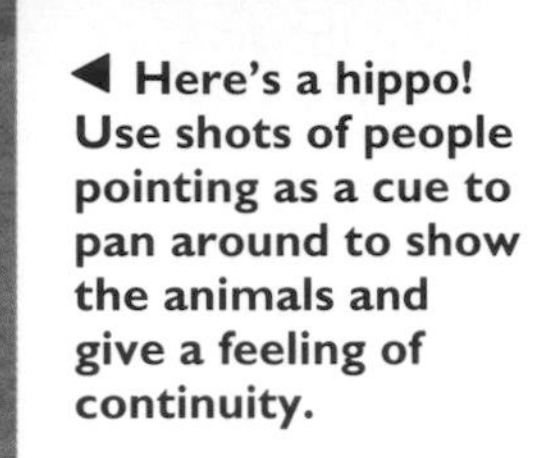

◀ **Here's a hippo! Use shots of people pointing as a cue to pan around to show the animals and give a feeling of continuity.**

▼ **A perfect end to your video – the family, loaded down with souvenirs, happily strolling towards the exit.**

the video. For example, 20 seconds of a particular swimming penguin, followed by a few seconds of children watching around the edge of the pool, followed by a brief shot of a penguin waddling about on the rocks, will probably make a more entertaining sequence than five minutes of performing penguins.

THE END OF THE DAY

By now everyone may be feeling a mite weary, but don't put your camcorder back in its bag yet! Those last few sequences will make all the difference to your home video. Once everybody has seen everything they want to see, they may decide to head for the souvenir shop. There is usually enough light in most shops for you to use your camcorder (but remember to switch to a 'tungsten' setting if you have manual white balance). Shoot a short sequence of the items that catch your family's attention and intersperse these close-ups with low-angle shots of the children choosing their presents. Then move outside for a concluding shot of everyone emerging from the shop, brandishing their purchases. Finish with a final shot of the family wandering, tired but happy, towards the exit of the park. □

WILD AND WOOLLY

Safari parks can present plenty of problems for videomakers:

- **BACKSIDE BLUES** – NEVER TRUST ANY ANIMAL TO PEFORM FOR THE CAMCORDER

- **MOVING PICTURES** – WATCH OUT FOR CATTLE GRIDS WHEN RECORDING IN MOVING VEHICLES – THEY CAN RUIN AN OTHERWISE BRILLIANT SHOT

- **BLACK BIRD** – THIS PARROT, SHOT AGAINST THE LIGHT SKY, HAS BECOME JUST A SILHOUETTE ON VIDEO. SWITCH ON THE BACKLIGHT FACILITY OR CHANGE POSITION

TAPES UNRAVELLED

ENCLOSED IN ITS BOX, VIDEOTAPE IS A VITAL PART OF EVERY CAMCORDER SYSTEM. THIS IS THE INSIDE STORY OF HOW IT WORKS AND THE TYPES AVAILABLE

The magnetic tape inside a video cassette is a highly sophisticated product. It is a plastic ribbon, a few hundred metres long, with a coating of magnetic oxide on one side. To avoid risk of pollution or damage it is enclosed in a plastic cassette. Its threading is done automatically inside the VCR or camcorder.

▲ **The camcorder tape housing opens automatically at a press of the Eject button. The opening mechanism is powered by the internal battery.**

Whenever you insert a video cassette into your machine, an intricate system of guides and poles pulls a loop of tape from the cassette and threads it around the video head drum. During recording the drum spins very fast and its two heads write 'magnetic signatures' on to the tape in microscopic tracks at an angle to the ribbon itself. The tape is pulled along by a capstan, and gradually wound on to a take-up spool as recording progresses.

In replay mode the tape is again wound around the head drum, but now the video heads are switched to 'read' mode. The magnetic patterns stored on the tape induce tiny electric

◀ **A Hi-8 camcorder with its compatible tape. These – like Video-8 – are comparable in size and shape to audio cassettes.**

▼ **A VHS-C tape with camcorder. This compact format will need a special adaptor to be played back in**

currents in the whizzing heads; they are amplified, processed and turned into the electrical signals a TV set needs to make pictures and sound.

All tapes work in this way, though they come in many sizes, shapes and grades. In Video-8 machines the sound is recorded with the vision signal, while some VHS camcorders 'write' the sound-track separately along the top edge of the ribbon.

TAPE FORMATS

The most popular camcorder format is Video-8. There are two basic types of tape, standard and Hi-8; the latter is only for use in the more expensive Hi-8 camcorders. In VHS format there are two sizes of cassette: VHS full-size (rare in camcorders); and VHS-C (compact). If your camcorder is a high-band model (S-VHS) you'll need the expensive S-VHS type.

Having made sure that the cassette size and type are right for your camcorder, you will often get two more options – running time and grade. In terms of running time, it generally pays to get the longest available in VHS-C and Video-8, and to go for the three-hour (E-180) type in standard VHS format.

Each videotape brand is available in various grades: Standard, High (HG), Extra High (XHG) and Pro, the price increasing proportionally. The more expensive grades wear better and are definitely worth considering if you are going to copy or edit from one tape to another. Pro tapes are the most strongly constructed and come in thin plastic cases.

VALUE FOR MONEY

When shopping for tapes you will encounter many makes. The cheapest is not always the best buy! A good guideline is to buy well known brand-names which you have seen on audio tapes, on camcorders or VCRs.

When abroad, be wary of counterfeit cassettes bearing lookalike logos of famous makes. If you buy tapes in Japan, USA and Canada you will find their running time does not correspond to that on the labels – European camcorders run at a slightly different speed.

Some types of tape can be difficult to get in certain areas of the world, so take plenty with you, especially as you can usually get discounts on ten or more cassettes at home.

When shooting, time passes quickly, and it is a good plan to take three times as much tape as you think you will need. Always keep tapes in their protective boxes when they are not in the camcorder. If there is any slack in the tape ribbon, take it up before loading the cassette in the camera – with a finger for VHS and VHS-C, with a pen or pencil in the spool socket for Video 8 types.

An often-asked question concerns the effect on tape of the X-ray security-check equipment at airports. This has no effect on videotapes or the programmes stored on them, nor does it affect the camcorder.

▲ A VHS, or full-size format tape with camcorder. The tape can be played back directly in a home VHS VCR.

As long as video cassettes are stored and handled carefully, tape faults are rare. Old, worn or low-quality tapes may cause 'drop-out' on the picture; the effect is of spasmodic blips and flashes across the television screen.

Screen snow

Tapes can get creased, with the effect of picture break-up and coarse bands of white dots and lines travelling slowly down the picture. Assuming the ribbon was not slack when loaded, creasing of tapes is usually caused by a faulty machine.

One of the main purposes of

ADAPTORS

If you own a VHS-C camcorder, an adaptor to fit your tape directly into the compatible VCR is a neat alternative to lead connections. The battery-operated adaptor is easy to use: you flip up the lid and slot the cassette into its niche (see below). You will then hear a whirring sound as the internal mechanism standardizes the VHS-C tape for VCR use. Video-8 tapes can be played back using separate Video-8 VCRs.

SOUND ADVICE

KEEPING UP THE QUALITY

In non-hi-fi VHS camcorders (those which do not incorporate stereo) the sound quality is geared to tape speed. Avoid LP (Long Play) mode if you want to capture the full audio frequency range, and avoid tape hiss. In hi-fi/stereo camcorders, and with all Video-8 machines, the tape speed has no perceptible effect on sound quality, though pictures are always better in SP (Standard Play), regardless of format or type.

videography is to save memories for posterity. Stored correctly, there is no reason why videos should not last many decades. 'Archive' tape is best recorded in SP mode for compatibility with future generations of camcorders and VCRs. Fully rewind the tape before storage; wound on the spool it will be safe from dust and damp.

STORING FOR THE FUTURE

Correct storage conditions are vital to preserve tapes. Stack cassettes upright in their boxes, flap-first. Keep them in a dry, dust-free environment where the temperature remains reasonably constant. If you are in any doubt about the storage environment, seal the boxed cassettes in air-tight plastic bags, enclosing a preservative silica-gel tablet or crystal-bag.

If you send tapes through the post make sure you pack them in padded bags, avoiding the type in which shredded fibre is used as a cushion — it can pollute tape. The type that uses air-bubble plastic for protection is best for video cassettes.

Whether storing or sending videotapes, label them clearly with date, subject, place and title: it will help to clarify things when you go through them 20 years on! □

TAPE CONSTRUCTION

Video tape consists of three or four layers, each performing a special function. The magnetic coating captures and stores pictures. The primer treatment binds the magnetic coating to the base film, while the base film prevents tape stretch and shrinkage. Pro tapes have a further back coating, an anti-static barrier.

MUSIC AND COPYRIGHT

MOST MUSIC - WHETHER YOU HEAR IT ON THE RADIO, BUY A RECORD OR GO TO A CONCERT - IS PROTECTED BY COPYRIGHT LAWS. BY CHOOSING A PIECE OF MUSIC FOR YOUR SOUND-TRACK YOU RUN THE RISK OF BREAKING THE LAW. HOW ARE YOU TO FIND YOUR WAY THROUGH THIS LEGAL MINEFIELD?

Copyright laws, recently updated in Britain in 1988, were drawn up to protect the rights of creative artists. Sound recordings, films, photographs, broadcasts on TV and radio, books, plays and music are all covered by these laws. So by dubbing a piece of music, be it Mozart or Michael Jackson, on to your home video you are technically infringing the laws of copyright. Even when you video at a wedding or party, for example, where music is played you are also committing an offence if you do not have clearance to record that music.

Although it is unlikely that you would be prosecuted over a genuine home video, made purely for domestic viewing, any kind of public showing, whether it is at a local library or video competition, could lead to a prosecution, whether the audience paid to see the film or not.

There are many different tapes of copyright-free music available commercially, but if you want to use music by a living, or recently dead, composer you will need to clear two copyrights; the copyright of the song itself, (via the Mechanical Copyright Protection

▼ Even amateur performances of music are protected by copyright laws – so check with the MCPS before you make your video.

Society – MCPS), and the copyright on the recording (via Phonographic Performance Ltd – PPL). The MCPS can be contacted at Elgar House, 41 Streatham High Road, London, SW16 1ER, tel: 081-769 4400, and the PPL at Ganton House, 14–22 Ganton Street, London W1V 1LB, tel: 071-437 0311.

If the author of a piece has been dead longer than fifty years, the work will be copyright free. The recording itself, however, is covered by the law. So, if you want to use a Schubert song, you will need to clear it only with PPL because the recording is in copyright though the music is in the public domain. Dealing with PPL is relatively straightforward. They offer a video dubbing licence, valid for a year, which costs £21.15 (incl. VAT) and allows you to use copyright sound recordings on not more than four individual video tapes.

◀ **The music of great classical composers, like Mozart, is no longer covered by the copyright laws – though they must have been dead for more than 50 years. But today this music is performed by other artists and their recordings are protected by the law and clearance must be given.**

MARY EVANS

Actuality and post-production

The MCPS draw a distinction between 'actuality' music (music performed during the shooting, at a wedding, bar mitzvah etc), and 'post-production' music (music that you dub on to your video at the editing stage). For 'actuality' music their licence costs £5.00 (incl. VAT) and allows you to record at one private function, and produce up to 25 copies of the tape. It also gives you copyright clearance on all the musical works which might be performed, (hymns, celebratory music, etc.) but if the works happen to be recordings, as opposed to live, you will also need the PPL licence. At the moment, the MCPS do not have a similar licence for the copyright of 'post production' musical works, and these need to be cleared individually, via the MCPS. This can be time-consuming, and the royalties may be expensive, so make initial enquiries with the MCPS Licensing Negotiation Department. Also check with them if you want to video a live piece, be it a school concert or amateur musical. The MCPS are planning to introduce a licence that will simplify the clearance of musical works for dubbing, although it is expected to be a bit more expensive than the 'actuality' music licence

An alternative is to become a member of the Institute of Amateur Cinematographers. Bona fide amateur film and video makers can then apply for the relevant licences via the IAC. Full membership costs £20, and associate membership is £16. For further details, you should contact The Administrative Secretary, IAC, 24c West Street, Epsom, Surrey, KT18 7RJ, tel: 0372 739672. □

▶ **You might want Michael Jackson on your sound-track, but remember: copyright clearance can be costly and complex.**

REDFERNS

Sound recording

PHOTOGRAPHS BY JO BOURNE

SOUND IS AS MUCH A PART OF YOUR RECORDINGS AS THE PICTURES. UNDERSTANDING THE CAPABILITIES OF YOUR BUILT-IN MIC WILL GO A LONG WAY TOWARDS IMPROVING YOUR HOME VIDEOS

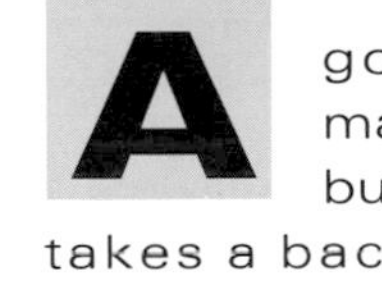

A good sound-track can make or break your video, but all too often the sound takes a back seat to the images. Learning to use sound properly is a vital part of using any camcorder with a little forethought it can add immensely to the quality and power of your shots.

Built-in camcorder microphones record sound on the tape as you shoot. While it is possible to edit the sound-track of your video later, the wiser course is to get to grips with the capabilities of your camcorder microphone, avoiding irritating sound problems from the start.

Although no two camcorder microphones are exactly the same, nearly all have what is known as Automatic Level Control (ALC), which means that the sound-recording level

LEARN HOW TO ***Listen and judge the level of ambient noise*** ***Test your built-in mic for sound pick-up*** ***Create a 'sound symphony' on tape***

◀▲ **WHAT THE LENS SEES**
A big day out in the city: London's Trafalgar Square offers plenty of opportunity for a good travel video sequence.

◀▼ **WHAT THE MIC HEARS**
You may hear your companion's commentary, but the mic will pick up all the ambient noise: traffic, repair work, tourists, fountains. Move the camcorder as close as you possibly can to your friend.

is automatically adjusted. But how does your camcorder know just which sounds to make louder, and which to make quieter?

BALANCING ACT

The way the Automatic Level Control works is that all loud sounds are made quieter and all soft sounds are made louder – it is thus a sort of sound balancing device. A problem can occur with this system when there is a sudden loud sound. For example, you might be standing under a tree videoing a tranquil country scene, complete with birdsong, when a bird-scarer suddenly goes off a few yards away. Because this sound is loud and close to the microphone the sound level will drop abruptly. The camcorder may take a little while (half a second to a second) to recover and 'turn' the background sound up again.

Some camcorders are built with a headphone (or Audio Monitor) socket so that you can listen to the actual sound you are recording. This is extremely useful if you are worried about wind noise or traffic rumble. If your camcorder does not have this facility, however, there are several methods you can employ to get better sound-tracks for your videos.

WANTED SOUND

A microphone, of course, does not hear in the same way as we do and cannot differentiate between the wanted and unwanted sound as we can. Suppose you are sitting talking to a friend in a crowded restaurant – all around everybody else is chatting, crockery and glasses are clinking, and there is music in the background.

Despite all this background noise you can pick out the voice you want to hear because your brain has decided which is the 'wanted' sound and turned it up. A microphone cannot do this – it simply picks up all the sounds around and mixes them together at the same level.

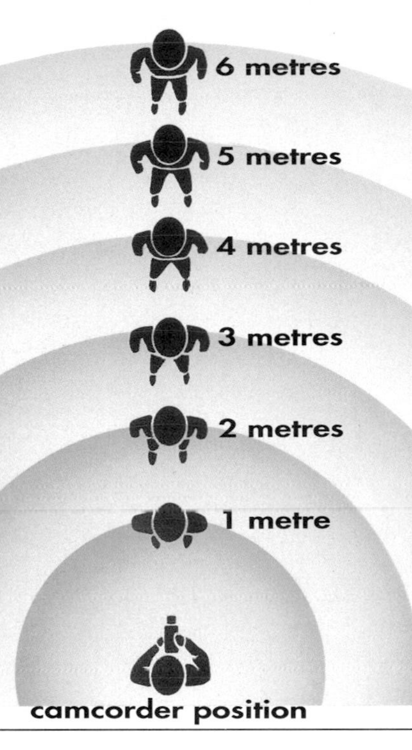

THE LAW OF SQUARES
Sound intensity decreases with distance according to the Law of Squares. If the source moves twice as far from the mic, the sound recorded will be a quarter of the intensity; if it moves four times as far, the intensity will be a sixteenth of its original strength. The ALC will generally compensate for these changes.

◀▲ DISTANCE TESTING
Ask your friend to walk away from you backwards, speaking in a clear voice all the time. You'll find that the voice remains constant to a certain distance with the help of the Automatic Level Control, after which it will become level with any background noise.

When you play the results of your recording back through a speaker, all the sound comes from a single direction and your brain will be unable to separate them. Your otherwise successful video has been ruined by a poor sound-track.

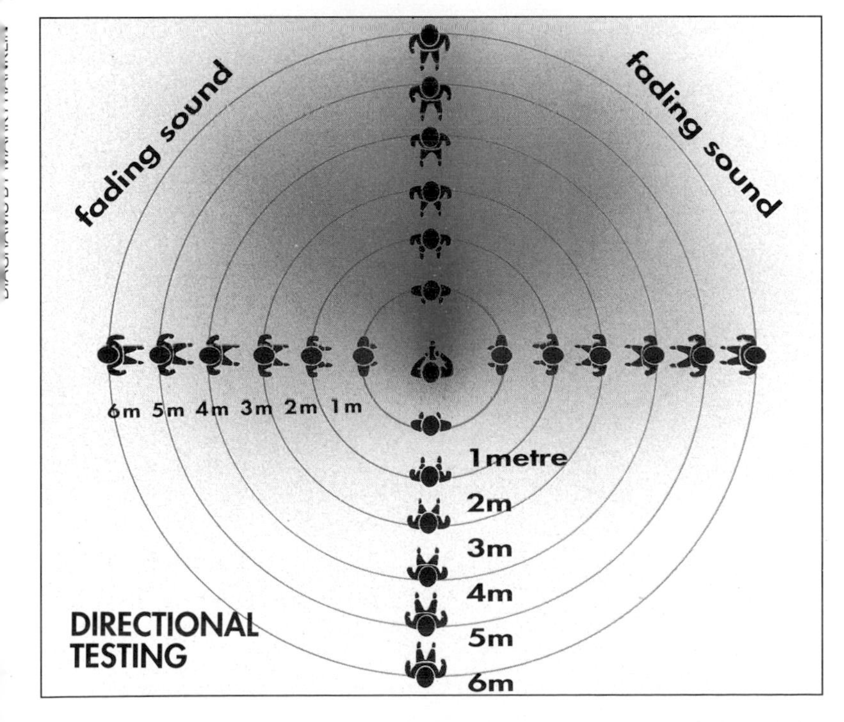

◀ ALL-ROUND SOUND
Test the camcorder mic's directional ability in the same manner. Most camcorder mics are omni-directional, which means they will pick up sound from all around. However, sound to the side and sound behind will generally be fainter than sound in front. Built-in mics have different capabilities depending on camcorder make and model.

TESTING, TESTING

The first step to take if you want to avoid such problems is to get to know how your microphone works. To do this, carry out the following simple experiments and write down the results. (This will save you making a lot of mistakes later. Remember to keep a notebook with you all the time if a particular effect worked, or something was a bit disappointing, you can always refer to it next time you are in that tricky situation.)

First you need an assistant with a clear speaking voice. Then set up your camcorder outdoors in as quiet a place as possible, a park or perhaps a large garden. Ask your assistant to stand directly in front of the camcorder, about one metre away.

Do not worry about framing or focus at this stage. Start the camcorder recording and ask your friend to say in a clear, level voice: 'I am standing one metre away from the camcorder and directly in front of it'.

Repeat this experiment with your friend in position directly in front of the camcorder at two metres ('I am standing two metres away. . .', etc.), three, four, five and six metres. This exercise will help you to define how sensitive your microphone is.

EVER INCREASING CIRCLES

Ask your assistant to stand one metre away, again directly in front of the lens and point the camcorder at them. From now on do not move the position of the camcorder at all. Start recording and ask your assistant to say clearly : 'For this experiment I am standing directly in front of the camera at one metre.' Get your assistant to walk round you, saying, 'I am now at the right side of the camcorder at right angles to it; I am now walking towards the back of the camera; I have now reached the back at one metre away. Now I am continuing to the left side of the camcorder and have reached the left

PHOTOGRAPHS BY LYNDON PARKER

TALKING NOTEBOOK

The built-in microphone makes an ideal 'talking notebook' if you are planning to write or overdub a new sound-track to your video at a later stage. Alternatively, the microphone may be used to provide a live commentary, mentioning names of places and people as you shoot your video, with either yourself or a companion acting as the commentator.

IMPROVE YOUR SKILLS

Kitchen Sink Drama

Combining pictures and sound in an arresting way can create excitement out of the most humdrum situations – such as someone doing the washing-up . . .

▲ OPENING SHOT
Our sequence begins with a silent close-up, held for a few seconds, of the washing-up bowl. Press Pause.

▲ CRASH!
Suddenly, as if from nowhere, knives and forks (thrown by your assistant) crash into the bowl. Press Pause.

▶ SQUELCH!
A loud squirt of washing-up liquid splashes on to the cutlery. Press Pause.

◀ **SPLASH!**
Change your shot to bring in the tap as water cascades into the bowl. Press Pause.

▶ **MUSIC!**
A hand mysteriously reaches out to turn on the radio – loud rock music begins to play as you slowly zoom out from the scene . . .

◀ **THE END**
Your zoom shot reveals your smiling assistant at the sink. Hold this shot until a convenient break in the music. When you replay the sequence on your TV you should have an arresting combination of rapid-fire sounds and pictures, sandwiched between your long-held beginning and end shots.

hand side at right angles to the lens; I am continuing walking until I am directly in front of the camcorder again.' Repeat this at two, three, four, five and six metres.

Play your tape back on TV at normal volume, and listen to the sound. Draw a diagram of what you did and mark on it what worked best. Did you notice that as your friend got further away the background sounds got louder? Were they acceptable, or was there a point at which the sound became unusable? What happened when your assistant walked round to the side and the back?

Now you have built up a complete picture of how your microphone picks up sounds from different directions and distances. These

Q

I tend to shoot a lot of wildlife videos. How can I get a good, clear sound-track of birds and animals?

Record as much as you can with the built-in mic, but if the sound is poor, cheat! There are many sound-effects LPs and CDs on the market, which you can use to dub over your original sound-track.

should be written down and used to help you get the best sounds when you are shooting real action.

It is very important that you get used to listening to sounds in the same way as the microphone. You need to 'switch off' the selective way you naturally hear sound.

For another experiment, take your camcorder and assistant to a busy location (a railway station or café). Now write down in your notebook where you are, describing the scene as accurately as possible. You will need it when you shoot action in these locations in the future. Write down what you think you can hear, everything – don't cheat! Then carry

◀ What seemed like little more than a gentle breeze at the time, may sound like a howling gale on your video sound-track.

▲ The quick solution to wind noise on a built-in mic – the Rycote Mini-Windjammer. A scaled-down version of what the professionals use, the Windjammer comes in all sizes to protect camcorder microphones.

GONE WITH THE WIND

The sound of the wind blowing across the face of the microphone is extremely difficult to avoid. If you have to shoot in windy conditions try not to move the camcorder about – this will make the noise worse. Some machines have wind reduction switches, or special built-in circuits. To test yours out experiment in windy conditions – ferry boats or theme park rides are ideal.

out the same experiment as you did earlier for distance and direction. (Remember that results will vary depending on the acoustics of your interior location.)

Common problems

The closer you get your microphone to the subject, the better your sound will be. Avoid zooming where sound might be a problem, just move the camcorder closer. Alternatively, you can ask your subject to speak up! .

Traffic noise and the sound of the sea are easier to control with simple planning. Either arrange the shot so that the camcorder is not pointing directly at the noise source or, better, remember that sounds decrease in level with distance. Move your subjects as far away as possible from the background, then zoom in a little, which will make the background appear nearer.

Lastly, don't forget that the built-in mic may pick up motor noise from the camera's zoom and autofocus, and handling noise – perhaps from the operating controls or a sudden change in grip while recording. □

NOISES OFF!
Think sound as well as vision and you will be able to avoid the most common problems associated with the built-in mic.

• **Flying low** – Filming in a rural setting, it's easy to overlook the shattering effect of a low-flying jet

• **Party pooper** – If you want to hear more than babble, move closer

• **Off the track** – From this distance, all you will pick up is ambient noise

Taking the lead

Video editing can entangle you in a veritable jungle of leads, cables and connections if you don't know your Scart socket from your Mini-8 plug

▲ **Camcorder makers have established their own plug/socket types over the years. If you plan to move into editing you will need to understand the different types of connection and what they are used for.**

Once your camcorder has captured all the wonderful sights and sounds you want on videotape, the next task is to replay them through your television set to see them in full colour. And, if you feel they could be improved by a little cutting and re-arranging, you might want to try your hand at some editing.

Before you start, you need to link your equipment together, and in order to make even the simplest connection, you need to acquaint yourself with all their plugs, leads and sockets. This may seem straightforward, but when it comes to hooking up, many camcorder users run into problems; there are so many different kinds of plugs and cables that you can end up trying to link your machines with a tangle of useless spaghetti.

The reason that so many people have difficulties with leads is that since the early development of audio-visual equipment, manufacturers, singly or in groups, have established and maintained their own individual plug and socket types. Although wide standardization is in progress, it has yet to filter down to much of the domestic machinery still in use.

Plugs and sockets

An understanding of what the different plugs and sockets are will go a long way towards helping you play back and edit your recordings. To begin with, take a look at your camcorder. Somewhere on its body you'll find an AV (Audio Visual) output socket. This is the starting point for your hook-up. Most camcorders have a Mini-8 pin socket, of

ILLUSTRATIONS BY ANDREW HARRIS

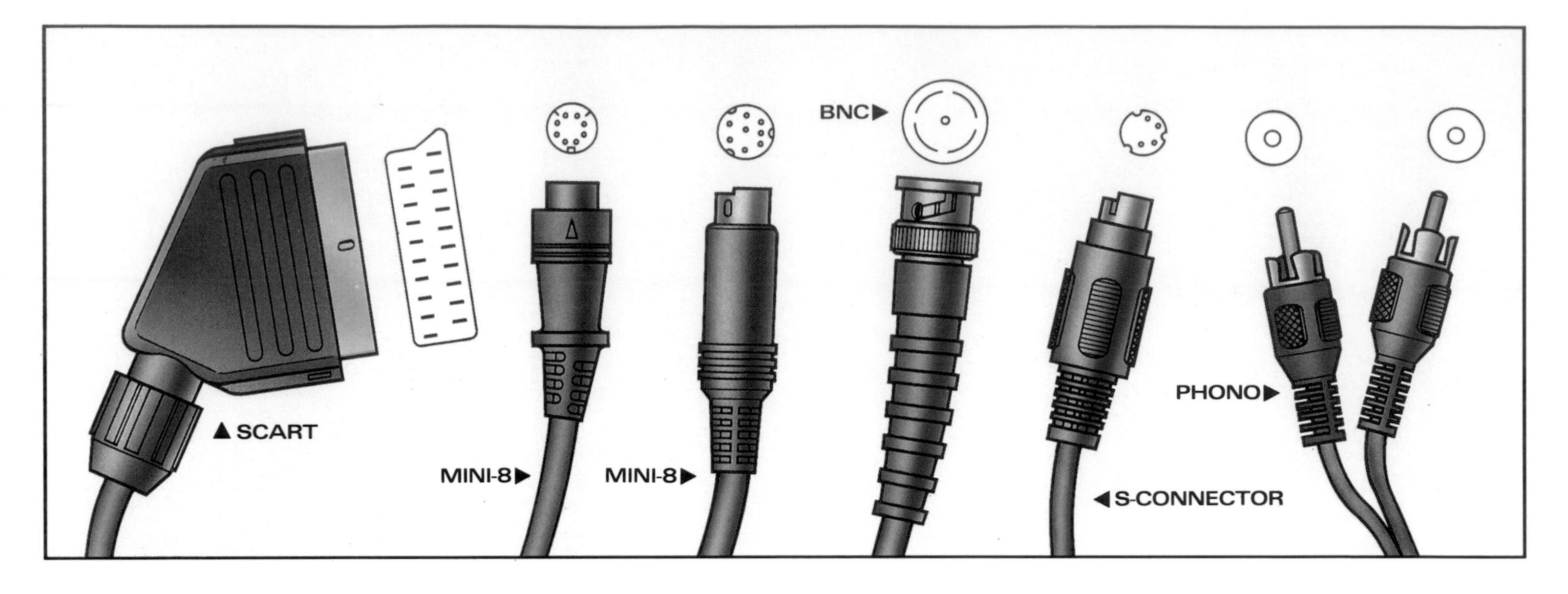

▲ Though the connectors vary, for most purposes all you will need is a lead with a Mini-8 plug on one end and a Scart plug at the other.

▼ If you are unsure which leads are best to use to hook up your equipment, consult your local dealer.

which there are three types in use. (Some manufacturers – notably Sony and Canon – provide separate phono sockets for audio and video.)

Next, take a look at the leads supplied with your camcorder. The camcorder end of your AV lead will have a matching plug. With phono plugs, ensure that you put them into the correct sockets – they are colour-coded for ease of use. Hi-band camcorders (S-VHS, Hi-8) have an additional 4-pin socket. This is called an S-terminal.

At the other end of the cable you need to have a plug which matches the input socket of your TV or video machine. Modern machines are manufactured with a 21-pin connector called a Scart socket (other names for this are Peritel and Euroconnector, and it is this which is being used to standardize equipment). For most purposes, a lead with a Mini-8 plug on one end and a Scart on the other is sufficient.

However, some TV and VCR models have different types of AV socket. You may encounter BNC or phono sockets for sound or even a 7- or 8-pin DIN socket for sound and vision. If the leads you want are not supplied with your camcorder, dealers will happily meet your needs. Likewise, if the lead you have is not long enough, you can buy quite long extension cables, typically 1.5 or 2 metres long, with a male connector on one end and a female connector on the other. However, in general the best idea is to keep all leads as short as possible, as they act like resistors, losing a small proportion of the signal as it travels through.

Extra TV channel

Many camcorders come with a little box of tricks called an RF (Radio Frequency) converter or modulator. It plugs into the camera's AV socket and has a 'flying' lead terminated in a TV aerial plug. In effect it is a flea-power TV transmitter working on about channel 36 – you can tune a spare button on the TV or VCR to it, and treat it just like another TV station. However, if you can avoid using the RF converter by hooking direct into

SOUND ADVICE

THE STEREO EFFECT

Camcorders (especially Hi-8) have excellent sound quality, but it's very easy to lose it in the cable connection system. RF convertors cannot carry stereo, and may even impair mono sound quality. If your TV or VCR is kitted up for stereo, maintain the link by using AV leads from a stereo camcorder. If you have a hi-fi system near the TV, connect its AUX-in terminals to the audio-out port of the VCR (eg. two phono leads) to bring a stereo video tape to vibrant life.

AV sockets, do. Direct AV connection gives higher quality picture and sound, with no risk of patterning interference or distortion.

S-VHS and Hi-8 camcorders have a second output terminal which is known as an S-connector. If your TV and/or VCR has a matching S-connector, link them together for the best possible picture reproduction – you will get higher definition and avoid 'cross-colour' patterning in areas of fine detail on the TV screen.

Recent large-screen TV sets have generally been developed with these S-connectors, but if your set boasts a Scart connector only, you can get the benefit of S-linking by using an RGB converter – several models are available.

With S-connectors a separate link is required for sound, in the form of an extra cable from the camcorder to the TV or VCR. Details are given in the user instruction books.

MAKING THE CONNECTION

The simplest hook-up scenario is the replay of the camcorder into the TV set:

- Connect the lead from the camera to the TV's input socket.
- Select AV, AUX or put the camcorder into play, and you're set.
- When shooting within range of the television, you can use it as a monitor during recording, but keep the sound turned down to avoid feedback or 'howl round'. A small, battery-powered TV or monitor can be employed as a colour viewfinder on location with the same hook-up arrangement, or via the RF converter.

In order to copy from your original videotape on to another tape:

- Connect the camcorder's AV output socket to the VCR's input port.
- Select AV, AUX or channel 0 on the VCR.
- Set the camcorder to play mode, and the VCR to record.
- You can monitor what's happening to the

▲ The invaluable European Scart socket/plug was designed as a standard connection system combining both audio and video potential.

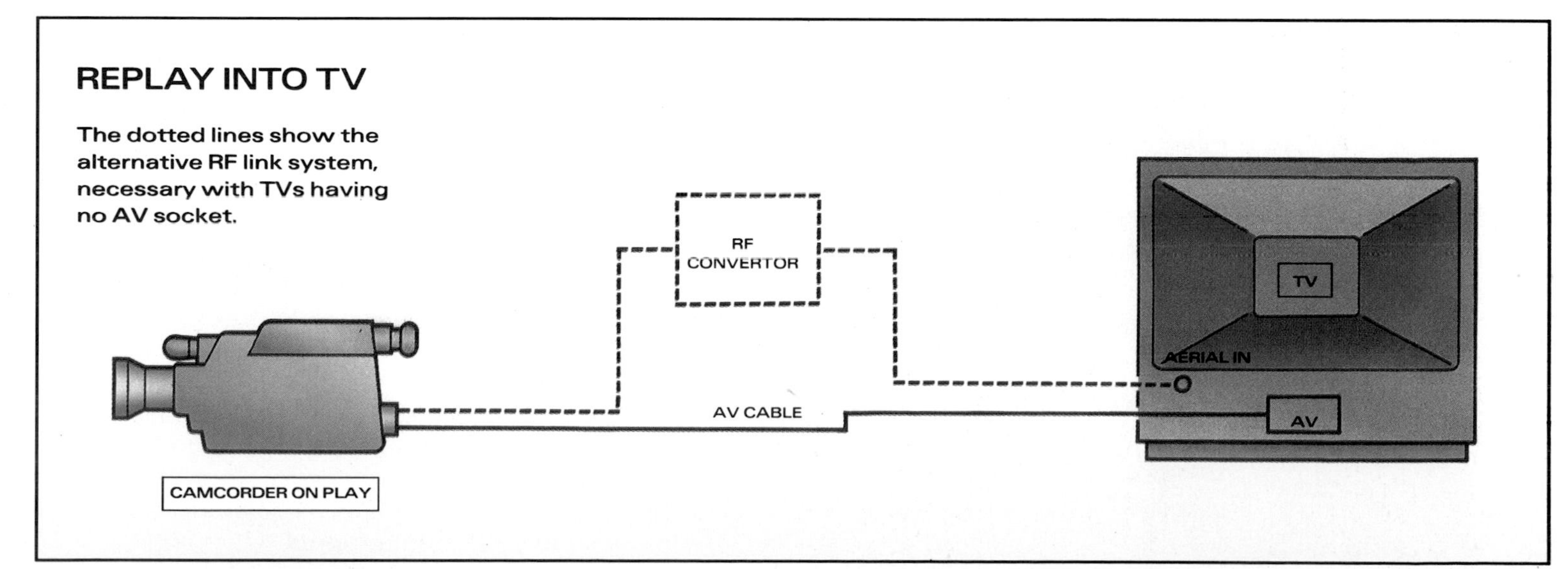

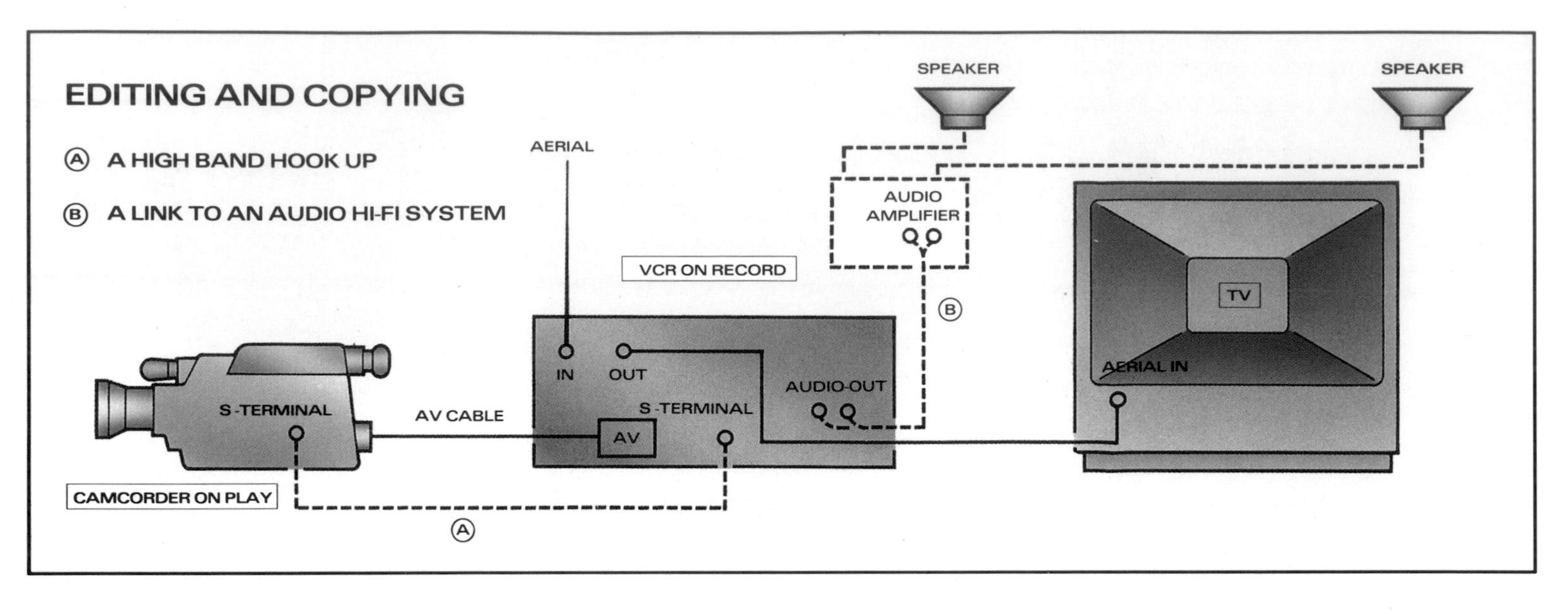

▲ **By connecting your camcorder to the VCR and TV, you can monitor your original recording on the TV and build up a master by using the VCR's Rec/Pause key.**

sound and vision on the TV which is already connected to the VCR.

To edit the tape, use the VCR's record-pause key to build up the programme you want on the 'edit master' tape. If the VCR is a high-band (generally S-VHS type) model, use the S-connnectors to take full advantage of the high-quality you have paid for.

A wide range of sound and vision accessories is available, such as mixers, enhancers, title generators, effects units and even computer genlocks which synchronize the pictures. All these are wired between the camcorder and the copy/edit VCR, and will come with the correct connecting leads and specific wiring instructions. The same applies to the sophisticated video editing systems that are now available.

To avoid confusion when linking home-video gear together, it is a good idea to regard the camcorder as a signal source; the TV set as a destination and the homebase VCR as either, depending on whether you wish it to record or play back.

CABLE CARE

Leads and connections need to be cared for:

- **Keep connectors clean to prevent oxidization.**
- **Never pull connections out of the sockets by the lead.**
- **Keep connectors off the floor where they may get trodden on.**
- **Never bend the leads as this may cause internal breaks. When not in use, coil leads into loose loops.**
- **Don't force connectors into sockets if they don't fit easily – you may bend a pin.**
- **Keep leads apart as some will act as antennae and pick up hum.**

The leads you need

If the RF converter or connecting lead you need did not come with the camcorder, you can probably buy it, bubble-packed, in a TV/video or hi-fi store. Check carefully that it has the plug terminations that your equipment needs. If you are still unsure of exactly what type of connections you require, take the various parts to the shop for advice. You will find that most shop staff are helpful and knowledgeable – let them save you time and worry! If the connection is not stocked, order it as an accessory from the agent or dealer for the camcorder manufacturer, quoting the model numbers of the equipment and plug type required.

Some special leads are not available from stock, in which case the dealer may be able to put you in touch with a specialist firm which can track down or custom-make a lead to your requirements. □

The Big Match

EVERY FOOTBALLER DREAMS OF APPEARING ON TV. NOW YOU CAN MAKE THOSE DREAMS COME TRUE, AND TURN YOUR TEAM INTO SUPERSTARS FOR A DAY

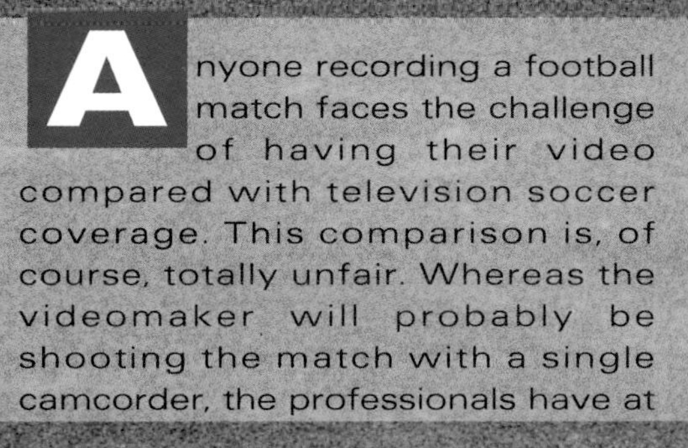

Anyone recording a football match faces the challenge of having their video compared with television soccer coverage. This comparison is, of course, totally unfair. Whereas the videomaker will probably be shooting the match with a single camcorder, the professionals have at least six cameras positioned at key places around the ground to ensure the best possible view. TV coverage is also boosted by action replays and commentators, but none of this should put you off: the important thing is to learn from the experience of top broadcasters. Start by studying television coverage, noting

PHOTOGRAPHS: JOHN SNOWDON.

▲ A close-up of your team's logo (using your camcorder's Macro setting, if appropriate) makes a good opening title.

PLANNING CHECKLIST

- **Take a tripod or other support, batteries, tapes and something to stand on**
- **Check touchline positions – halfway line, penalty area, and behind goals**
- **Rehearse panning and zooming shots**
- **Practice using manual focus directly behind one of the goals**

▲ A mid shot of your team running on the field introduces all the players. You could name each one, using the built-in mic, as they pass out of shot.

not only where the cameras are positioned, but how they are used and the sort of shots they get.

A soccer match is one occasion when some sort of camera support is almost essential. A sturdy tripod with a good fluid head for smooth camera movements would be ideal. If you are in a crowd, a monopod could serve you well, and if you are shooting hand-held, a chest pod or shoulder holster will give stability and support during long takes.

Power source

With a 90-minute match to record, sufficient battery power is vital, especially as motorized functions such as autofocus and zoom increase the rate of power consumption. A battery belt would be the ideal power source to record the whole match.

You will also have to bear in mind tape length. VHS-C and S-VHS-C tapes run only 45 minutes, so would be insufficient to cover the whole of one half if there were any delays or injury time. Consider using Long Play

▲ During the warm-up, you may be able to get on the field for a few action shots.

▶ This moment of calm before battle commences gives you a chance for a wide shot of the captains and one of the most important people on the field – the referee.

(LP) mode to double recording time. One way of lessening these difficulties could be to go for 'edited highlights' of the football match, concentrating on showing your team to best advantage.

You are thus more likely to have plenty of power in reserve to shoot entertaining moments of the build-up to the match. As kick-off time nears, Try to get on the field to record your team running on to the pitch and warming up. Don't forget to include brief shots of the referee and linesmen, who are sure to play crucial roles in the course of the match.

Camera positions

Whatever kind of match you are covering, a key camera position will be close to the centre line. Use this position to give a wide shot of about one third of the pitch. You can then pan smoothly to follow the action as it moves from one end of the pitch to the other, revealing the flow and strategy of the game. By staying wide, your movements will be gentle

▶ Best positions are (1) at halfway line; (2) between the edge of the area and the touchline; and (3) behind a goal.

CAMERA SENSE

WHICH WAY ARE THEY PLAYING ?

If you are shooting from different positions, remember not to disorientate the viewer by 'crossing the line'. Shoot everything from one side of an imaginary line drawn between the centre of the two goal posts. If you shoot from both sides of the pitch, viewers will be confused about the direction of play.

ILLUSTRATION BY MARK FRANKLIN

and for much of the time the camera can be static, with the game simply unfolding in front of you.

As action develops, particularly around the goalmouths, you can tighten the shot by zooming in. Don't try going too tight; you will have problems following the action (finishing up with a series of rapid, hard-to-view, panning movements going back and forth) as well as focusing. Instead, work on the principle of going in tighter as action develops, but coming out wide again as soon as you feel your camera movements are becoming too complex or rapid. By both panning and zooming at the same time, you will create a smooth, flowing effect.

Q ***What should I watch out for when videoing a match played under floodlights?***

The effects of floodlights on your camcorder's auto-iris. Shots taken against lights will cause it to close down. Avoid low angles and adopt high shooting positions. **A**

Another good camera position will be level with the edge of the penalty area, ideally from a high position. This is excellent for covering set plays, such as free kicks and corners, on your side and that end of the pitch. You will have to accept, however that much of the action is likely to occur many metres away from this vantage point and that your coverage of the game will be intermittent. The important thing is to watch the match action carefully and start recording as soon as you can predict that the ball and players are moving in your direction – remembering that because of 'back spacing' you will lose the first couple of seconds of action after you press Record.

Shooting from the penalty-area position requires you to adopt the same principle used for shooting at the centre line, but here you can expect to get a closer view of some of the crucial moments of action, such as when one of the players gets in a shot at goal.

Providing your camcorder has a manual zoom control, you can take advantage of this facility for corners and throw-ins. Start close in on the corner post with the figure seen full-length and zoom out rapidly as the kick is taken, panning across at the same time to follow the direction of the ball towards the goal. Motorized zooms do not operate quickly enough to accommodate this kind of

◀ Make sure you are in position for set plays such as this corner sequence showing the big centre back powering a header into the top corner.

▶ Half-time is an ideal opportunity for unguarded close-ups as the team relax.

CAMERA SENSE

FOLLOW THE ACTION

The key to action camera work lies in predicting events. Shoot your video with both eyes open if possible so you can scan the area not shown by the viewfinder. Keep the framing wide to reduce any movement you have to make and leave plenty of 'walking room' for players to move into.

▲ A penalty or a fair tackle? Freeze-framing when your video is played back on TV later will settle arguments.

►▼ A brief pause in the action for an injury provides a contrast to the ebb and flow of the game. Shots of the 'crowd' provide ideal cutaways between action sequences.

shot, so instead, frame up a wider shot of the corner kick that just requires you to pan with the trajectory of the ball to bring the goal area into shot. In this way you will not miss any of the vitally important goalmouth action.

LINKING SEQUENCES

Because this particular position will not provide you with continuous coverage, you should be aware of how each of the sections that you shoot will edit together in-camera. Try to ensure that you start and end each section of recording with the camera static, so that you do not end up with ugly edits with two pans cut together in the middle of their movements. You can also avoid 'jump cuts' by varying the front and end composition of each shot – for instance by starting with a tightish telephoto shot of the action and widening as the players move towards you. Alternatively, simply shoot a cutaway of the onlookers between each recording of the match progress. Vary the cutaways between wide shots, long shots of groups and close-ups of individuals.

Another vital camera position is next to one of the goals. Place yourself in a spot where you can see

◀ **This sequence taken from a position just to one side of goal is a good example of framing a head-on shot and allowing the action to unfold for a static camera.**

▶▼ **Northern Town's star centre forward leaves the last defender floundering as he bears down on goal – ensure you have a comfortable shooting position to maintain your shot for as long as the action lasts.**

the whole of the goalmouth in the viewfinder, either by choosing a position behind the goal, mid-way between the posts or at a three-quarters angle to them. Spread your tripod's legs fully to achieve a low angle – this will dramatize the action.

Goalmouth action

From this position you will only be able to record short sections of the game – but they are likely to be the most exciting and rewarding to shoot. Your own knowledge of the teams playing will tell you which end will probably see most attacks on goal; you can switch ends at half-time to an equivalent position if the balance of the game is in one direction.

From this low angle, and with the lens in its widest position, it should only be necessary to pan and tilt to follow goalmouth action. With the camcorder positioned behind the goal, mid-way between the posts, you will be ideally placed to cover corner kicks from either side. Simply pan just as the kick is taken, following the ball into the goalmouth.

In this low camera position, you should watch out for problems with exposure. If there is too much sky in shot because of the angle, the auto-iris will close down and the players will be seen only in silhouette. You can overcome this problem by cranking the tripod up a bit higher (so that not much more than a third of the the sky is in the viewfinder) or by using manual iris control or Back Light Compensation (BLC), if your camcorder has one.

In the shadows

Keeping the lens angle at its widest and shooting in full daylight provides maximum depth of field and minimizes focusing problems. But the autofocus system may have trouble with fast-moving players running past the camcorder, or be fooled into focusing on the goal net, rather than the action beyond. Try using manual focus, if your camcorder permits.

Room to move

The advantage of videoing club matches is that you can adopt any position around the pitch and be sure of getting coverage of some aspect of the game. Your freedom of movement will undoubtedly be a determining factor on how you shoot. For a small club event you may well be able to record different parts of the match from each of the identified positions, moving between them as the action unfolds. For the most challenging and potentially exciting shots, get close to the touchlines and shoot at shoulder height. Use this position to follow individual players. Aim to frame the full length of the figure with enough space around him to include the ball and be prepared to pull out to a wider angle as passes are made or the player moves in your direction.

▲▶ A penalty is a vital moment in any game. A position directly behind the goal may be best, but you will need to focus manually or the autofocus may fix on the net. The less spectacular view from the touchline ensures that focus problems are avoided.

ELEVATED SHOTS

When shooting from the touchline, try to use an elevated position that allows a view down over the pitch. This angle eliminates foreground and helps to give viewers a reference point by showing the pitch markings. Improvise a platform by standing on a box or lightweight stepladder. Alternatively extend your tripod's legs and central pedestal and stand on a box, stepladder or chair to operate.

In order to follow the action you will often need to pan, tilt and zoom at the same time. These can be tricky shots, but they will produce rewarding moments. Remember that focus can also be a challenge with players constantly on the move, and using manual controls (including the zoom) requires practice.

Change of pace

Break up action sequences from pitch-side positions with cutaways of match officials or players not directly involved in the action. These moments of calm will serve to emphasize the hectic pace of the rest of your video.

The course of a football match is, of course, extremely hard to predict, and you will inevitably find yourself

▲ **Get the team together for a photo at the end. Ask them to sing a burst of the team song as a finale to your video. Hold your shot until they have finished.**

SOUND ADVICE

BE YOUR OWN COMMENTATOR

With a school or club match you can even provide your own commentary as you go along. Headphones are available with microphones attached for this very purpose or more simply you or a friend can provide commentary with the aid of the built-in microphone. Alternatively, your friend can use a separate, auxiliary, hand-held mic and stand nearby as you shoot.

▼ **At the final whistle, join the players as they celebrate on the pitch.**

shooting some passages of play that turn out to be less interesting than others or are spoiled by poor camera work. These can be eliminated by later editing, but decisions made at the time make for a better result.

Remember to make a note of the footage counter at the end of each sequence; if the next one does not come up to your expectations, you can simply rewind and re-record over that section of tape.

When the final whistle goes, you may be able to join your team on the pitch. Ask the players to pose for a group portrait to comemmorate the occasion. A long-held shot such as this will make a fitting end to your video of the Big Match.

OUT-TAKES!

SICK AS A PARROT!
Expect groans from armchair 'experts' if your coverage contains these errors . . .

• **END TO END** – DON'T WASTE POWER ON ACTION TAKING PLACE TOO FAR AWAY – MOVE TO A CENTRAL POSITION

• **SPOT THE BALL** – FOLLOWING THE BALL INTO THE SKY IS SURE TO MAKE YOUR AUTO-IRIS CLOSE UP, RESULTING IN SILHOUETTE IMAGES

• **POOR ANTICIPATION** – TRY TO KEEP BOTH YOUR EYES OPEN AS YOU SHOOT TO KEEP UP WITH THE PLAY

BATTERIES

BATTERIES GIVE CAMCORDER USERS FREEDOM TO SHOOT WHEN AND WHERE THEY CHOOSE. YET THESE APPARENTLY SIMPLE CHUNKS OF STORED ELECTRICITY NEED CAREFUL TREATMENT

▲▼ A Nicad battery in position, and a standard battery charger. The charger can also provide a source of direct power when plugged into the camcorder.

A camcorder battery is the portable power supply you tack on to the machine when you want to be fully mobile as you video. To the unpractised eye, they all look much the same. But 'power pack' batteries do vary in size and capacity, and some are better for camcorder use than others. To find the right one for your camcorder, and to get the best out of it, some knowledge of how a battery works can be helpful.

THE BATH PRINCIPLE

Imagine a bath full of water. If you remove the plug, water flows out under pressure until it has completely drained away; the bath now needs refilling. A battery works in exactly the same way. When you buy a battery it is full of power (measured in watts). When you use the battery it is drained of electrical current (measured in amps); and the pressure put upon it as it drains is measured in volts. Once the battery is empty, it needs refilling (recharging). Rechargeable batteries (like NP47s, NP77s, and NP31Rs) are the obvious choice for camcorder use: they are convenient and, if properly employed, economical.

When you go shopping for batteries, your first real worry will concern type and size. Of the main battery types – lead

acid and nickel cadmium celled – nickel cadmium (Nicads) are invariably the lighter and more handy: consequently they are ideal for camcorders.

Secondly, you should be aware of your camcorder's specific voltage. Camcorders need a supply at either 6v, 9v, or 12v; check what your model requires against the voltage marked on the battery.

Life before recharge

But above all, perhaps, you will want to know how long the battery will last before it goes flat. To give some indication of its life before recharge, a battery is marked in mAH – the measure of how many milliAmps (mA) of current can be supplied per hour (H=hours) may be worked out by multiplying the two figures together. Bear in mind that the larger its mAH, the bigger (and more expensive) the battery.

PHOTOGRAPHS BY LYNDON PARKER, DIAGRAM BY MARK FRANKLIN

▲▼ Tacking the battery on to the camcorder is extremely easy – especially when you know which features go with which.

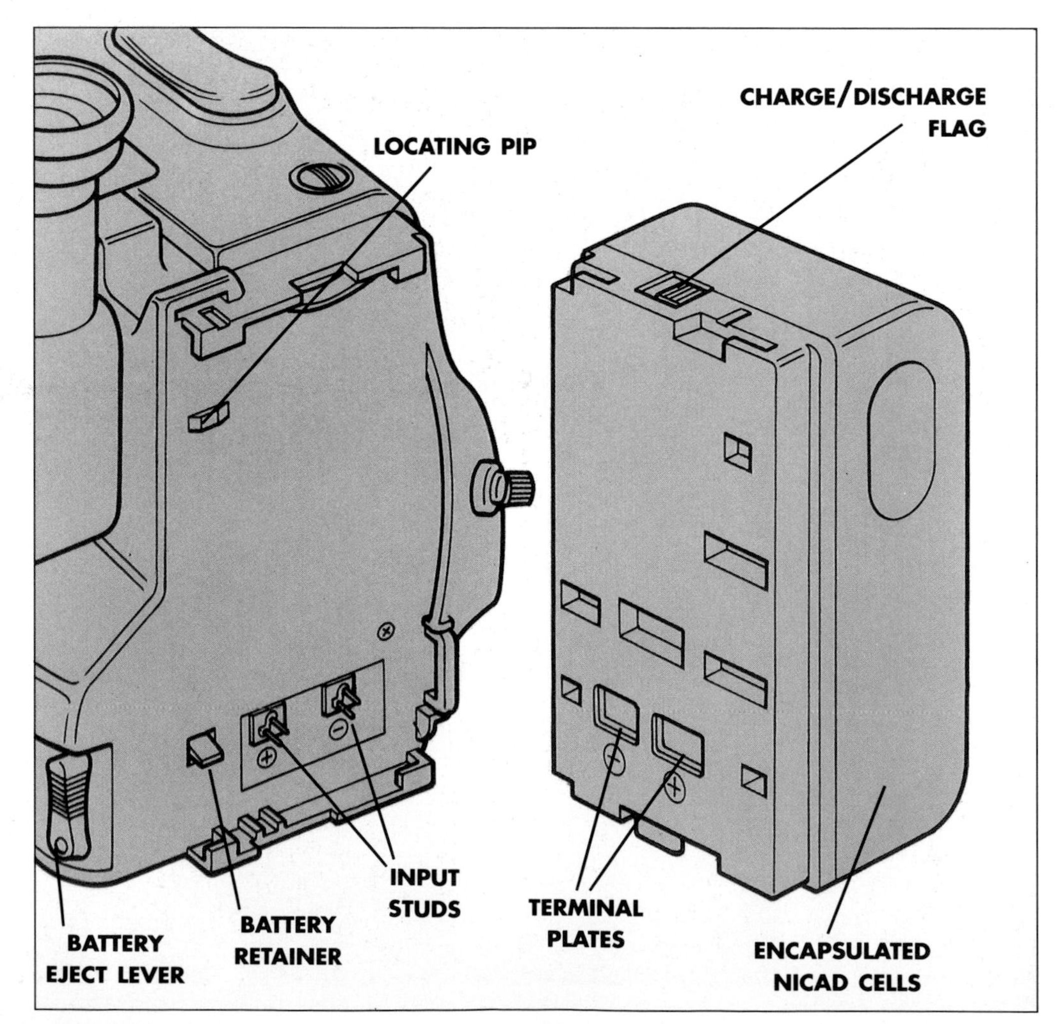

Solid-looking as they are, Nicad batteries need to be well treated to give you good service. To help you get the best from them, here are a few practical tips:

- If the battery gets too cold, the camcorder will cut out prematurely. If the battery gets too hot, it will rapidly go flat. In cold weather, keep spare batteries in your pocket; in warm weather, keep spares in a cool-bag. Swop batteries around every so often when you are filming.
- Dampness makes a battery go flat quickly, so wrap it in a soft cloth, not in a polythene bag, which will not prevent it from getting damp.
- Dirt reduces battery performance, so gently clean the contacts with a cloth before use.
- Batteries are designed to be used until they go flat. If the battery is recharged before it is fully discharged, it can 'remember' it only has to work for a short time before being recharged, and this becomes its normal – irritating – behaviour.
- The charging time for batteries is crucial. If a flat battery is designed to be charged for one hour, charging it for two will shorten its life. Follow the manufacturer's recommendation. □

Camcorder Travel

Some of the best home videos are made on holiday. Yet filming in a foreign country can present problems. Here is a general guide to trouble-free camcorder use abroad.

There are few better ways to use your camcorder than to record a holiday, but whether you are going abroad or staying in your own country, there are certain things that you must take into account when travelling with video equipment.

Obviously, much depends on your destination, how you are going to travel, and how committed you are to carrying all your equipment around with you. If you are going by car, or by car ferry, taking your camcorder and accessories should not prove a problem. Going by air brings with it the limitations of space and weight, but if you are planning to go cycling in the mountains you would have to be pretty dedicated to add a camcorder to your luggage.

Whichever way you plan to travel, you need to be well prepared, and the first essential is a protective carrying case. Hard cases offer the best protection but a soft case is much easier to carry on the shoulder. Whichever you choose it should be both shock-proof and water-resistant. Make sure that you choose one that will carry every thing you want to take with you.

Take as many tapes as you think you'll need. Similarly, pack extra batteries to give you plenty of power, bearing in mind that the autofocus, zoom and pause operations all take an extra toll on battery time.

Q ***Can I buy camcorder tapes abroad, or should I take plenty with me?***

Tapes are available in most countries, but they may be more expensive or poor quality (watch out for counterfeits). Tapes from North America and Japan have running times which don't correspond to those on the box, as the operating speed of European camcorders is slightly different. **A**

Check voltage and the socket configuration of the country you are visiting – you may require a special voltage transformer and plug adaptor.

A tripod may take up too much space in your bag; a monopod would make a good lightweight alternative.

Lights and additional mics are optional, but remember to pack a UV (Ultraviolet) filter to protect the lens, an ND (Neutral Density) filter to cope with excessive brightness, a blower brush, a cleaning cloth – and the manual. No matter how well you think you know your camcorder, a

ILLUSTRATIONS BY DAVID ASHFORD

quick read of the manual will remind you of functions you don't often use, such as title superimposers or caption generators, and may add to your final recording. (If your camcorder is new, take time to shoot a few test videos so that you are familiar with its workings).

Finally, all your equipment should be insured. It may be covered, up to a certain value, under home contents insurance – but make sure. For travel, you should have an all-risks policy (including 3rd party) which covers theft and damage, even if it is left in a car or hotel room.

Anyone who has ever seen the treatment handed out by baggage handlers across the world would shudder at the idea of leaving a valuable camcorder at their mercy. In addition, the luggage hold of a plane is not temperature-controlled as the cabin is, so any equipment packed in a suitcase is subject to extremely cold (and damaging) conditions.

Hand baggage

On the whole, then, it is a good idea to take your bag on the plane with you. (You may even want to record part of your flight – though it is best to check with the airline as safety regulations

CAMERA SENSE

ANYTHING TO DECLARE?

- **Camcorders, like many other 'luxury' goods bought abroad, are liable to import duty at British customs. Always take photocopies of receipts for equipment (especially if you will be passing through US or Japanese customs) and make a list of everything you have.**
- **The country you are visiting may require a customs declaration form and an import certificate. Check first.**
- **Some countries limit the amount of equipment you can take in; find out before you leave if this applies to you.**

▲ Some museums are happy to let visitors record the exhibits (some charge a fee), while others – such as the Louvre in Paris – forbid the use of camcorders inside the doors. Check in advance to avoid disappointment.

► Pack carefully, taking everything you think you might need, including receipts, spare batteries and tapes, and your manual.

▲ **Once off the aeroplane, you still have to negotiate customs. If stopped, be prepared to produce a detailed list of your equipment to avoid delay.**

may not allow it). Forget the notion that you can avoid having your camcorder and tapes X-rayed by packing them in your suitcase: on the contrary, they will be treated to a higher dose of X-ray than by the hand baggage vetting system.

There are two schools of thought on X-ray security. One is that they do no damage to either camcorders or tapes so there is no need to worry about losing the record of your holiday. The other suggests that while the camcorder itself is not at risk, the tapes may be, so to be on the safe side, you could ask airport security to hand check all your tapes.

Q ***Will I be able to play back my recordings on the hotel room television?***

In many countries, the TV system will be different to the British standard, which means that you will not be able to watch your recordings. If you don't have an electronic viewfinder on your camcorder, you could invest in a miniature colour monitor. A

Whichever option you choose, avoid taking your tapes through the metal detector or other electro-magnetic inspection devices.

Customs regulations vary from country to country, but it is worth taking receipts for (and a full list of) all your equipment to prove it was purchased here and to avoid paying a hefty import tax when you return.

Once you have arrived at your holiday destination you will probably be keen to start recording. However, there are some safety considerations worth bearing in mind. First, security. If you are staying in a hotel remember that hotel theft is not uncommon in holiday resorts. Rather than leave your camcorder in your room when you don't want to take it with you, ask if you can put it in the hotel safe. If you are travelling in a car, locking your equipment in the boot will not necessarily keep it safe. Resorts attract thieves and if you have been spotted using a camcorder and returning it to your car, you will be an obvious target.

RESTRICTED VIEWING

Keep your camcorder close to hand whenever possible; it will be safer and you will be prepared for any unexpected video opportunities. Keep a firm grip on it, too – camcorder snatching is not unheard of. Don't forget you are carrying a valuable piece of equipment which may have a high resale value.

Before shooting it is worth giving some thought to any restrictions on the types of subjects which you are

SPECTRUM COLOUR LIBRARY

COMSTOCK

▲ **Video religious shrines with sensitivity. Never disturb prayer, except with express permission.**

allowed to video. If you are in any doubt, your travel agent or tour guide will be able to give you an outline of the local etiquette. Restrictions on things you are allowed to film may include anything to do with national security, such as airfields, dockyards, sometimes even bridges. If in doubt, don't film military installations – people have been sent to prison for years for making this mistake!

SHOWING RESPECT

When in a church or any place of religious worship, it is customary to ask permission to film the interior. Indeed, many churches and cathedrals demand this. A fee is charged or a voluntary donation made in most cases. Be sure not to intrude upon those using the place for prayer; in many churches photography of any kind is forbidden during services anyway.

Similar rules apply to stately homes and other buildings; it is only courteous to ask permission of the owner or someone in authority beforehand. If asked not to film, do not try and do so 'undercover'. This may involve having your equipment confiscated or broken, or even a brush with the authorities, none of which makes for a happy holiday.

It is important to bear in mind the cultural differences you will find when travelling abroad. In certain countries – either because of poverty or religious taboo – the local inhabitants may object to being filmed. Find out before you start. You will probably find that most people are delighted to be filmed (if you ask them politely), especially children.

SERVICE CHARGE

In some countries, it is the custom to give a tip to anyone you want to video. Check when and how much is appropriate with your tour operator, guide or local hotelier. Tipping like this is not automatic by any means, and thrusting money at what may seem to you to be impoverished villagers may cause great offence.

It would also be wise to check with your travel agent as to whether there is a limit on the amount of equipment you can take with you. Some places do have restrictions, and it is not worth packing your equipment only to have it confiscated at customs.

It may be worth taking a note of place names and dates as you go; by the time you view your tape at home they may be difficult to recall. Just a little careful preparation will give you both a successful holiday video and a lasting memory of your trip. □

CAMERA SENSE

PROTECT YOUR EQUIPMENT

- **Sand and water will damage your camcorder. Take a waterproof case if you go to the beach.**
- **Avoid extreme heat. Never leave your camcorder in direct sunlight or in a hot car.**
- **Keep the camcorder in a shock-proof and water resistant carrying case when not in use.**
- **A UV filter fitted to the lens will protect it from scratches, and will add clarity to sea and sky shots.**

PHOTOGRAPHS BY LYNDON PARKER

Sharp focusing

AUTOFOCUS IS EASY – UNTIL YOU ENCOUNTER REFLECTIVE WATER OR WIRE MESH. A QUICK SWITCH TO MANUAL CAN IMPROVE YOUR VIDEOS ENORMOUSLY, AND GIVE YOU MORE SCOPE FOR CREATIVITY

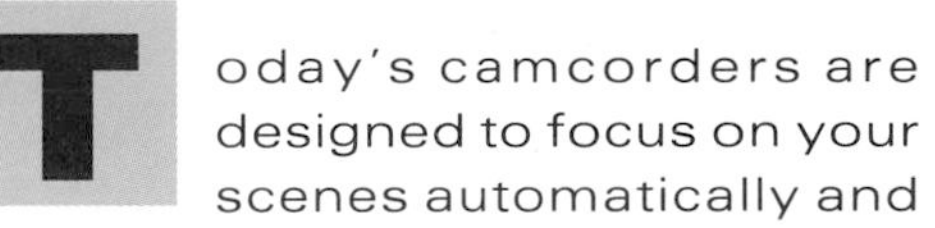

Today's camcorders are designed to focus on your scenes automatically and manually. The Autofocus mode, which videomakers are happy to use most of the time, is a miracle of modern technology and is capable of handling most situations. Designed to focus sharply in the centre of frame, it is excellent for basic shooting, but it can sometimes go awry, leaving you with fuzzy images or the 'wrong' object sharp in your picture.

The first problem the videomaker will encounter with Autofocus systems is that they have a tendency to 'hunt' a sharp zone, and constantly re-adjust focus as you move the camcorder or the subject moves. (You have probably noticed how the focus ring turns as you shoot.) This 'hunting' can be irritating, because the

LEARN HOW TO ▶ ***Anticipate shooting problems with Autofocus*** ▶ ***Focus manually on particular occasions*** ▶ ***Use Manual focus for creative effect***

▲ REFLECTIONS **Autofocus systems have problems with water. If the water is in motion, the system will have difficulty focusing. If the water is calm, the lens will focus on the surface only and not show what is underneath.**

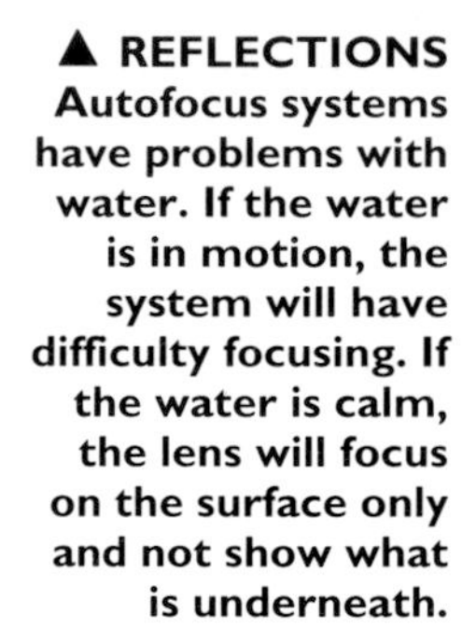

▼ Reflections from shiny cars can confuse the focus in the same way.

images move in and out of focus as the lens settles on a sharp zone in the picture, so it is worth familiarizing yourself with the 'search time' of your machine so that you can anticipate problems before they arise.

Two systems

There are two principal types of Autofocus system. The first, the infrared system, measures the distance of the subject in the centre of the frame by bouncing an invisible light beam from a small window beside the lens to the subject, then back to the camcorder, which sets the lens using a Servo motor. This is why some camcorders are confused when they encounter glass, water and netting.

More sophisticated systems use TTL (Through-The-Lens) Autofocus, which 'senses' the sharpness of the image with a microchip in the viewfinder. Such systems compare the relative sharpness of areas of the picture to set the 'best' focus. This average focus ensures maximum sharpness throughout the picture.

Many camcorders are fitted with AE (Auto-Exposure) systems for specialist applications such as portrait and sports photography. AE settings alter the shutter speed and aperture to maximize sharpness for slow-motion replay in Sports mode or to reduce the 'sharp zone' in Portrait mode so that the subject is separated from the background. Some AE systems also alter the responsiveness and speed of Autofocus. Choose the option that suits you – and the event – best.

No available Autofocus system offers full creative control. You may, for example, want a sharp image at the side of the frame or to shoot a

▲ CROWDS **People walking in front of your lens can temporarily throw the Autofocus out, while with too many people at close range, the camcorder will not know who to focus on at all. Switch to Manual on wide-angle for good, sharp results.**

QUADRANT

city skyline at sunset through trees. Ordinarily this is not possible; however, some camcorders have a focus-lock that allows you to select the focus on any given subject in the viewfinder and over-ride Autofocus as long as you keep the focus lock on. All you have to do is place your chosen subject in the centre of the frame, set the lock, then re-frame to shoot.

DIFFICULT SUBJECTS

Autofocus is baffled by situations where two subjects at different distances overlap. If you are filming an animal in a cage, for example, and you cannot safely or easily move the camcorder to a position between the bars (or zoom in between them), switch to Manual focus. With animals or birds behind netting, it is also best to move further away and focus with the lens at a longer focal length. The netting will then be so far out of focus that it will not obscure the subject.

Shooting in a crowd can send your Autofocus system into hyperdrive as there are so many moving subjects at varying distances – and you may want to pan – so the Autofocus will never find a stable subject to focus on. The solution is simple. Set the camcorder to Manual, and choose a wide-angle setting (which reduces camera shake). The wide angle gives you great depth of field, so you can move the camcorder freely and allow people to move back and forth without the lens 'hunting' wildly.

Autofocus systems can also be confused by other glossy, shiny surfaces such as car exteriors and mirrors, and even by surfaces which give little or no reflection, such as

TTL AUTOFOCUS

Incoming light is split by a prism, and passed to 24 sensors (each a micro-lens containing two photo-diodes), which examine the centre of the image. The information is analysed by a microprocessor, which gives 'drive lens in' and 'drive lens out' commands to the focus drive motor, depending on whether the focus plane is in front or behind the subject.

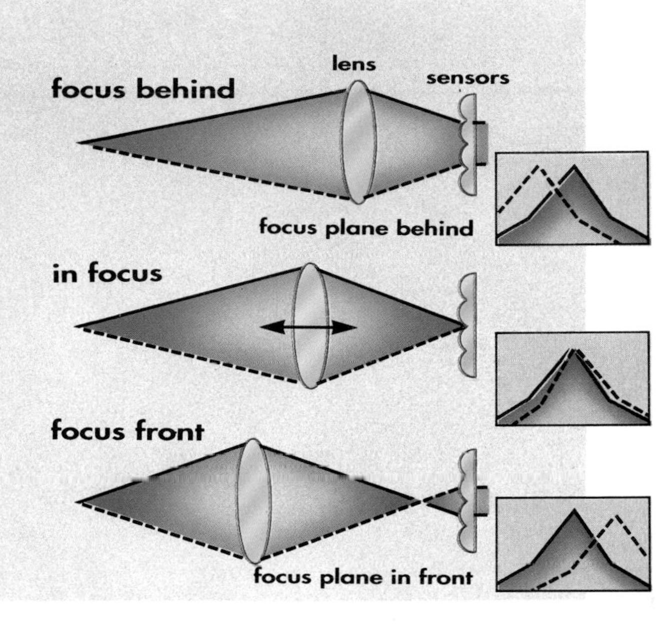

◀▼ SHIFT FOCUS
Use this Manual technique to add interest to your shots. Frame the picture, then twist the 'sharp zone' slowly away from you to reveal your chosen subject in its full glory.

dark clothing, hair or fur. In these instances, check your focus by moving to a wide-angle setting, then focus manually.

The final points to bear in mind are that Autofocus systems are less reliable in low light when there is no contrast, or when there is extreme contrast between the two halves of the screen. They can also be perplexed by backlit subjects. The remote-control handsets for TVs and stereos also use infra-red, which can interfere with the Autofocus, too.

MANUAL WORKS

Most camcorders can be switched from Autofocus to Manual focusing mode. Manual focusing techniques improve the creativity and quality of

Q ***What is the best way to focus sharply on a subject seen in a mirror?***

Manually set the focus on the image, and not on the surface of the mirror. This will be the distance from your camcorder to the mirror and back to the subject. This procedure should give you a sharp image. **A**

your videos immensely and offer you the freedom to video the scene as you want. You can shift the 'sharp zone' towards the camcorder by turning the focus ring clockwise, and away from the camcorder by turning it anti-clockwise.

Remember that the sharpness of the image is affected by focal length and aperture. A telephoto shot has less depth of field than a wide-angle shot taken closer. The less light there is the wider the aperture you need: bad light reduces your depth of field.

Selective focusing is a useful trick that emphasizes certain elements of your video image. You can change the focus manually so that elements

IMPROVE YOUR SKILLS

Shifting the Focus

Watch any TV programme and you will see how the professionals use focus to create emphasis and interest in their scenes. Take out a tripod and try it for yourself!

▲ A GLASS OF WINE?
Set the camcorder on a tripod and switch to Manual. Zoom in on the wine and glass, focusing on them alone to give real emphasis.

▲ CHEERS!
The pouring finished, pull back to include the person serving. Her companion is in the shot, but remains out of focus, because of her closeness to the camera.

▶ Now change positions to record her companion accepting the wine. The shot set-up is as before.

◀ **LOOK OVER THERE!**
Her attention is distracted by someone over her friend's shoulder.

▶ **HELLO MUM!**
Change your position again and take a telephoto shot of the boys waving in the distance.

◀ **COME AND JOIN US!**
Now pull the foreground into focus to reveal the whole scene. Everything is now in focus because of the width of the angle. With practice, adjusting the focus in such scenarios will become second nature, and you will be able to use Manual focus whenever you want to add drama to your videos.

of the scene are not sharp, and your main interest is very clear. Out-of-focus elements in the foreground can mask unwanted and distracting details, such as an ugly car park in an otherwise beautiful valley.

It is often the case in video that less is more, and with selective focusing you can suggest an environment without showing it. For example, give a 'cafe' feel to a sequence by placing unsharp silhouettes of a coffee pot, coffee cups, and a vase with a flower in the foreground of your set-up, then concentrate on the people, the real centre of interest.

SHIFT FOCUS

Another useful technique is that of shift focus, in which you change the focus from one object to another in the same shot. This trick is often used when two people are in the frame, one closer to the camera than the other. Shifting the focus between two subjects at different distances also shifts the emphasis from one character to another and tells a story.

For example, you want to shoot a golf ball being putted into the hole. You need a low angle at green level to give the shot impact. As the ball is

MOVING SUBJECTS

Autofocus systems cope with subjects moving towards or away from the camcorder in the centre of the frame as long as the subject virtually fills the frame and is not moving too fast. Fast-moving subjects mean you need to anticipate the action and pull focus manually if you want your images to remain sharp. Racing cars of all kinds make for good practice.

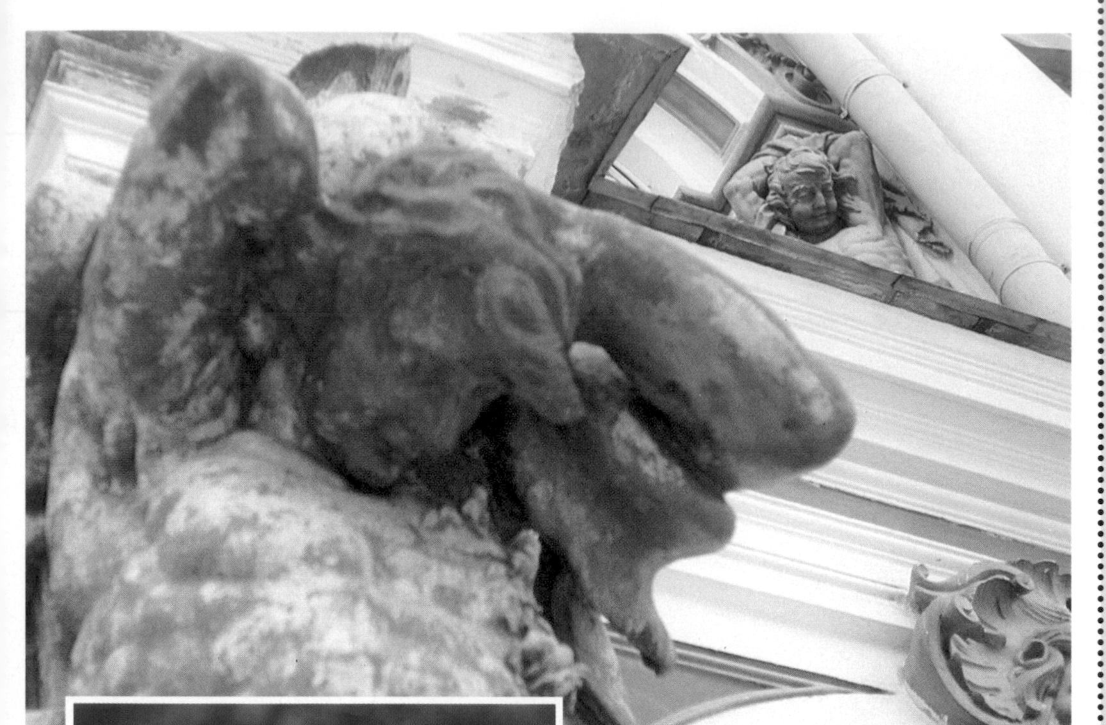

▲ **AUTO-LOCK FOCUSING**
Focus-lock on the statue in the upper right-hand corner, then move down to compose your shot.
◀ **To ensure that the rose is in focus, not the background, fix the auto-lock on the flowers. Then zoom out and compose your shot.**

putted towards you, shift focus so the ball stays sharp and the hole comes into focus.

There are three main techniques to successful manual focus.

- Prefocus: know where and when peak action occurs and let the subject enter the 'sharp zone' for maximum visual impact. The calibrations, marked in feet and metres on the lens, enable you to anticipate focus. This method is most useful at the telephoto end of the zoom.
- Follow focus: follow the action towards or away from the camera by moving the 'sharp zone' with it.
- Shift focus: being able to move the 'sharp zone' towards and away from the camcorder repeatedly during a single shot. This takes experience.

The time you spend familiarizing yourself with lens operations is time well-spent. Develop your techniques and ability to handle your camcorder in a variety of situations. □

CAMERA SENSE

MANUAL MAGIC

Set up a shot using a tripod. Zoom into a close-up on your main subject and focus manually until it is sharp. Re-frame your composition, bearing in mind you can now zoom in to a close-up without losing focus. With this technique, you can record two subjects at the same distance on either side of the frame and pan between them without altering focus.

OUT-TAKES!

FOCUS FAULTS

In Autofocus mode, some shots seem doomed to fail. Try to be alert for outside elements that could ruin your video . . .

• **Blowing in the wind** – On Autofocus, corn swaying in the breeze will cause the lens to 'hunt' for something to focus on

• **Out of the way!** – The unforeseen becomes the all-too-obviously seen when people walk in front of the lens, throwing your subjects out of focus

Assemble editing

RAW FOOTAGE IS USUALLY A RAGBAG OF GOOD AND BAD SEQUENCES. BASIC ASSEMBLE EDITING WILL ENABLE YOU TO CUT OUT THE PARTS YOU DON'T WANT, AND JOIN UP THE PARTS YOU DO

Once you have sat down and reviewed your uncut video at leisure, you will undoubtedly find much in your tapes that you would like to edit out. You may also want to vary the pace of your story, to alternate lively sequences of activity with slower 'atmosphere building' scenes, and to compress events which may in reality have contained much that was not very dramatic into a punchy selection of golden moments.

The next stage in the process is to consider exactly how to make those changes. There are several ways you can do this, but the most basic of post-production editing methods is assemble editing. The idea behind it is

▲ By cutting bad and boring sequences from your video using Assemble Editing techniques, you will finish with a film that tells its story more concisely – and doesn't stretch its audience's patience.

ILLUSTRATION BY EMMA WHITING/THE ART MARKET

very straightforward: one shot or sequence is assembled on to the end of the previous one, just as you might join a series of coaches to an engine to form a train. On a practical level, it involves simply copying material from your original tape on to a fresh blank tape, omitting from the copy tape any of your unwanted material.

During the copying process there will be some slight loss of quality in the sharpness of the image, the accuracy of the colours, and the clarity of the sound. However, these losses will be minimal if you use high-grade videotapes and high-quality equipment. Remember that the better the technical quality of the source material, the better the quality of the copy tape will be. (Some camcorders have an edit switch, which reduces the losses incurred by copying.)

No extras needed

You can assemble edit from your camcorder to a domestic VCR, as long as it has an edit facility with a flying erase head. This ensures 'clean' edits without electronic disturbance. Any combination of tape formats can be used – Hi-8 to VHS, S-VHS to VHS and so on. The only extra items of equipment you may need are audio and video leads (or, better, a multiway Scart cable) to send the video and audio signals from camcorder to VCR. Use the TV as a monitor for both machines.

Before starting the editing process, view your original tape several times, and make precise notes about which sections you want to include in your edited programme. Use the counter on your source machine to note the exact points where the good shots you plan to keep begin and end.

Start small

Having made your editing notes, you are ready to start. Don't be too ambitious at first. Start by assembling whole sequences; then, as you grow more skilful, try your hand at editing individual shots together. The exact details of assemble editing will vary according to the equipment you are using, but the general procedure is as follows. First, to

ASSEMBLE EDITING STEP BY STEP

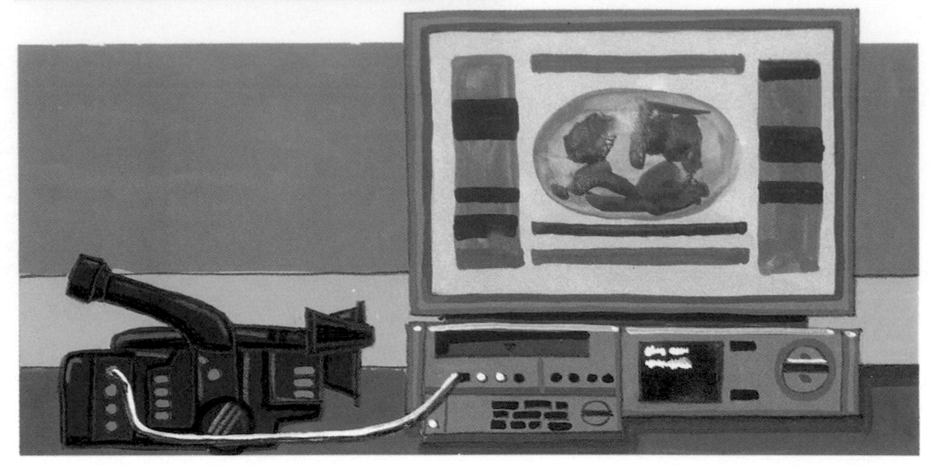

1 Basic set-up: camcorder connected to Edit VCR. Use the TV as a monitor for both.

4 Insert original master tape in camcorder. Find first edit 'in' point (using TV as monitor). Press Pause.

▲ Before assemble editing as above, remember to 'log' your tape by writing down the counter code numbers of your chosen shots or sequences. These will serve as rough edit guides.

CAMERA SENSE

EDITING 'ON THE FLY'

Some camcorders have no Pause button. To edit from these requires a little ingenuity. Find the 'in' point for the first shot. Rewind a few seconds. As your 'in' point approaches, release the Pause button on the VCR. Press Pause (or Stop) on the VCR when you reach the 'out' point. If you have gone to Stop with the VCR, play back the shot you have copied. Press Record/Pause at the 'out' point. Go back to the camcorder and find the 'in' point for the next shot.

2 Insert new Edit tape into VCR.

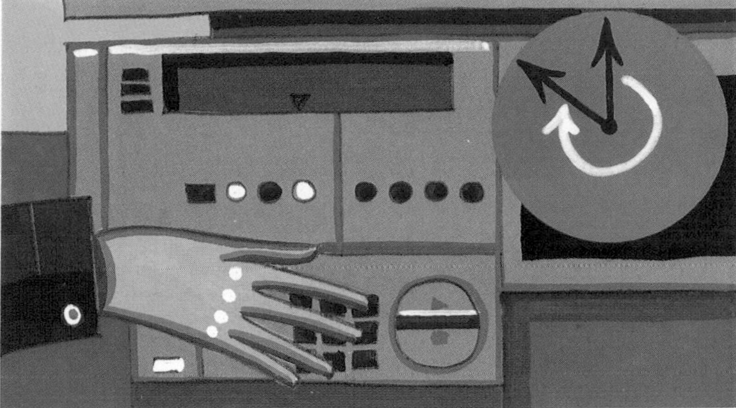

3 Play for 50 seconds. Press Record and Pause.

5 Release both Pause buttons simultaneously. Copying starts.

6 When you reach the 'out' point, press Pause or Stop on both machines to cease copying.

ensure a really stable edit at the the start of your programme, create 50 seconds of 'black' at the beginning of your new tape by recording with your camcorder in Video or Edit mode, and with the lens cap on, or iris closed. This leader at the start of the tape will also serve to protect the opening scenes of your programme against drop-outs, other glitches or possible damage that can occur over the beginning section of a tape.

Cueing up

Second, set the VCR to Record and Pause. Remove the recording tab from your original tape so that you cannot accidentally erase your source material, insert the tape into the source machine and cue up the sequence you want to transfer, by pausing the camcorder about one and a half seconds before

SOUND ADVICE

SMOOTH TRANSFERS

In assemble edit mode, pictures and sound are transferred automatically together, so the edit points need to be chosen carefully with sound as well as image in mind. To achieve a smooth effect, look for edit points where the camera was steady and settled, and where you will not be abruptly chopping off a word, a section of dialogue or a sound effect. Otherwise you may end up with a video with a good visual flow but a confusing sound-track.

▲ Assemble editing gives you shooting freedom. Sure in the knowledge that you can edit and join later on, you can shoot a subject from as many different angles, and for as long, as you wish.

EDIT FOR EFFECT

Basic assemble editing allows you to bring out the potential humour of a situation. Thus, for example, what seems to be a white elephant of a sequence showing a barbecue that obstinately refused to light properly, can be amusingly transformed. Simply edit together in quick succession the climaxes of the repeated failed attempts, and drop all the extraneous, in-between action.

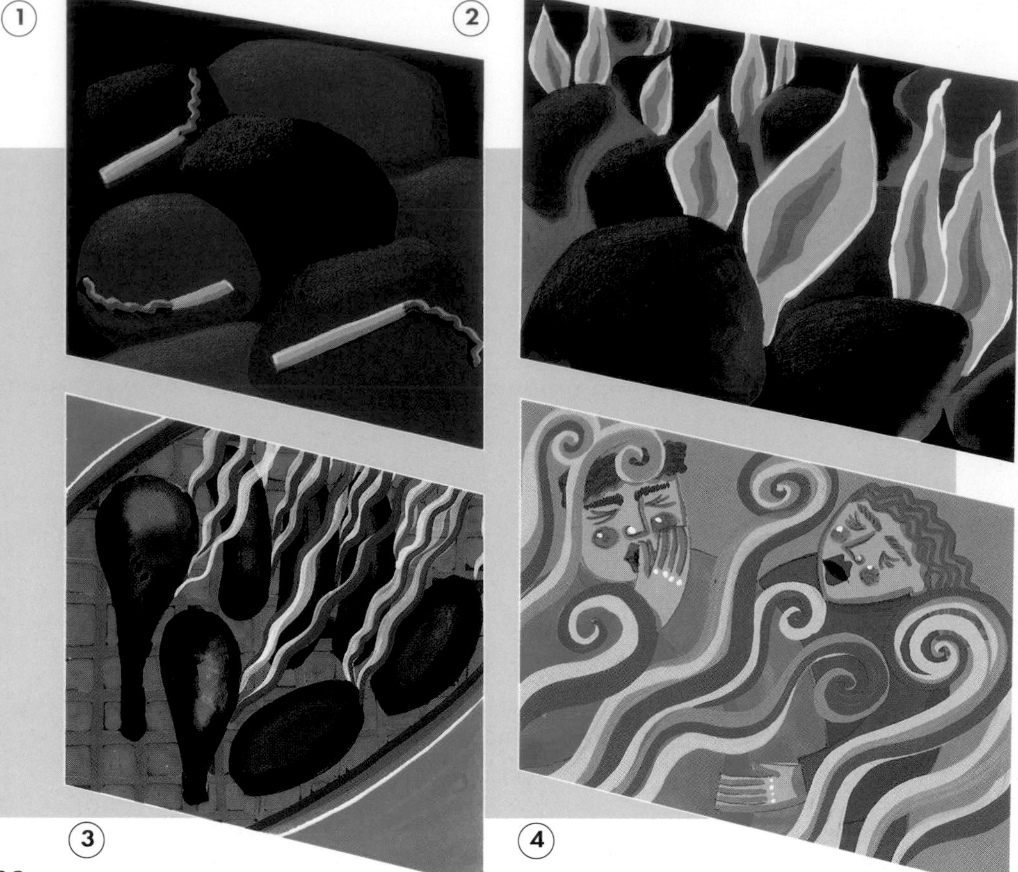

▲ These four shots represent the edited highlights of a 'chapter of accidents' at a barbecue. Shot 1 shows matches fizzling out. After lighter fuel was poured on the coals in a desperate attempt to get the barbecue going, shot 2 shows flames blazing upwards. In shot 3, the flames have died down, but the food has not been watched and has burned. In 4, fat from food has fallen on the coals – resulting in billowing clouds of smoke choking guests.

you want the edit to start (this allows the machine to search for an electronically compatible edit point). Having cued up the edit start points ('in' points) on both tapes, release the pause buttons on both machines simultaneously, and continue recording until you have gone past your selected end point, then stop both machines. Rewind and check that your edit occurred at the right point, then run the sequence through to your end point ('out' point) again. Put the VCR on Record/Pause while you cue up the next shot or sequence on your original tape. Again, release both pause buttons at the same time, and carry on as above. Although the procedure is very straightforward, it may seem rather clumsy and laborious, especially at first. However, with a bit of practice you will become adept at the fiddly business of cueing up the shots, and co-ordinating the button-pushing on both machines. Spend a bit of time laying out the equipment so that you can comfortably reach both sets of controls and see the monitor easily. If your machines have Variable Speed Search and Frame-by-Frame Advance facilities, you can greatly increase the accuracy of your edits.

Assemble editing is a good home-based editing method, but by virtue of being a linear, sequential process, does have its limitations. If you change your mind about the previous edit, and then decide that you want to redo it, you will lose a few seconds from the end of the shot, choosing a new, earlier edit point. (You cannot re-edit cleanly near the original edit point because the electronic control track is broken at the edit point.)

Q

What would be the ideal home set-up for assemble editing?

A VCR featuring front-mounted AV terminals, jog-shuttle control and audio-dub facilities, and, possibly, separate television monitors to allow you to cue up the edit points while comparing both images.

Your first attempts at assemble editing may sometimes be slow, painstaking and possibly frustrating. But the process really does repay persistence: once you have mastered its relatively simple techniques and seen the great improvements it can make to your videos in terms of flow and interest, you will be ready to move on into still more rewarding areas of editing practice. □

Down on the farm

A VISIT TO A FARMYARD, WITH ITS FASCINATING SIGHTS AND SOUNDS, MAKES A FUN DAY OUT FOR THE FAMILY – AND GIVES YOU A GREAT CHANCE TO PRACTISE YOUR VIDEOCRAFT

Never work with children or animals is a favourite maxim of professional performers. But for the family video-maker, the exact opposite is true. Children and animals together make excellent video subjects, and a farm visit in particular offers endless scope for variations on the theme – with the added attraction of all sorts of splendid agricultural machinery for those whose hearts don't melt at the sight of a frolicking new-born lamb.

FARM FAVOURITES

There are many sorts of farms open to the public. There are small city farms, designed to give very young children close contact with animals, and to allow the older ones the experience of helping to care for the animals. There are working farms, organic and otherwise, and there are farm parks which specialize in rare breeds. These are especially interesting as you can visit animals such as pot-bellied Vietnamese pigs, or traditional breeds no longer found on the modern

PHOTOGRAPHS BY MIKE HUNT

◀ **This sequence in long shot effectively introduces mother and child as they make their way across the farmyard.**

farm, such as magnificent Suffolk Punch horses. Obviously the kind of activities you can expect to see will vary according to the time of year and the type of farming involved. A mixed farm, on which crop growing is combined with animal raising, will offer most variety for the visitor.

Bear in mind that a working farm is a potentially dangerous place, so you will need at least one other adult on such a visit to supervise the children while you operate the camcorder.

LOCATION SHOT

You may want to shoot a title shot or sequence beforehand. This might be a shot of a large-scale map with the farm clearly located on it. You may need to set your camcorder on Macro to get details, and to tape the map to a wall, so that you can shoot from the tripod at an angle of 90 degrees. If your tripod has a centre-pole attachment you can spread the location map flat and shoot down on to a table or the floor.

You could try employing a very small camera movement at this point instead of a static shot, such as a tilt down from the top of the map to your chosen point. This will serve to reinforce the idea of movement and setting off on a journey. If your camcorder has a special facility for superimposing a title, this could go

PLANNING CHECKLIST

- **Before setting off, study and, if desired, shoot publicity material from the farm you plan to visit**
- **Remember to take spare, fully charged batteries and spare tapes**
- **Useful accessories to take might include a small clip-on 20-Watt video light, for shooting in dark sheds, and a monopod to help ensure stable shots**
- **When you first arrive, make a brief 'recce' of the farm to check its layout**
- **Practise any panning or tilting shots that will set the scene at the beginning of your video**

PHOTOGRAPH BY LYNDON PARKER

▶A rapid series of contrasting farm scenes – perhaps, as here, featuring your child, a tractor and a curious cow – could make an interesting section for the opening to your day on the farm.

A - PAN
B - STATIC SHOT
C - PAN
D STATIC SHOT
E - STATIC SHOT
F STATIC SHOT
G - PAN

ILLUSTRATION BY TREVOR LAWRENCE

over your map shot. Then fade or cut to the next scene.

Having selected your farm, try and find out as much as possible beforehand about what is on offer: is the farmer himself going to meet you and spend a little time with you? Is there a farm trail to follow, a fixed itinerary designed to lead you from one site of interest to the next? Or will you be able to roam around and explore on your own? Will the children be able to touch some of the animals? (They should not feed any animal unless given permission by the farmer.) Will there be certain things happening at specific times – for example, cows brought in for milking? Will there be special activities going on while you are there, such as sheep-dipping?

▶ Getting the attention of a farm beast is a sure-fire way of introducing a humorous interlude into the flow of your programme.

Roving reporters

Now is the time to give some thought to the style of your programme. You could create a video equivalent of a page from the family photo album. In this case, will you simply document events as they unfold, selecting the most expressive details to capture the feeling of the outing? Another possible approach would be to involve the children as roving reporters, doing little presentations to

▲▶ The more your human subjects interact with the animals on the farm, the more dynamic the videotape will be.

camera to move the action on from one scene to the next. They could also do mini-interviews with some of the farm workers.

Equipment check

Decide in advance whether you are going to take a tripod. Although it could be useful for some set-ups, it is probably better, unless you are exceptionally energetic, to stick to hand-holding your camcorder. You will also need to be able to respond quickly to things as they happen. Remember that you need to keep the zoom lens on wide-angle to increase the steadiness of hand-held shots. (You also get the benefit of increased depth of field.)

As with any important expedition, it is vitally important to check through all your equipment the day before, ensuring that all the batteries are fully charged and that you have plenty of tape. If you have not used your camcorder for some time, have a little practise to remind yourself of the precise layout of the controls. A hand-held camera operator needs the same sort of hand to eye co-ordination as a tennis player!

▶ Zooming out from the hands of the people to reveal their faces adds to the interest of the sequence. The intrusion of another cow into the shot makes the ideal point for a cut.

On arrival at the farm you have chosen, take some time to have a quick survey of the immediate surroundings before starting to shoot. This could be a good moment for an introductory piece to camera from your child reporter, perched on a gate perhaps, or standing in the middle of the farmyard. If the farm has an authentically rustic-looking entrance sign, this would also make a good 'found' title, with perhaps a verse or two of 'Old MacDonald' sung by the kids while you shoot.

Alternatively you could begin with some impressionistic close-up details. The children's welly-clad feet splashing through a puddle (watch

SUNLIGHT AND SHADE

On a sunny day a farmyard will offer an extreme range of lighting values – from the full sun of the open yard to the semi-shade of partly covered pens, to the deep shade of interior sties and stables. These variations in light will cause constant fluctuations in the automatic iris of the camcorder, if you pan from light areas to shade, or vice versa. Even within a static shot, if you include part of a bright exterior in a predominantly shady shot, the contrast may be too much for the video image to handle, with the brightly lit portions being burnt out, or the darker ones underexposed. Careful framing can avoid many of these situations.

out for water on the lens!), the old sheepdog wagging his tail in welcome, a hen enjoying a dustbath.

Follow these details with a wide shot of the farmyard to give a sense of the layout of the place. Take time to choose your angle – there may be some steps you can climb, or a gate you can sit on to get a good view.

VARIED SHOTS

You may find that a single static shot will provide a good sense of space; however, a panning shot is more likely to reveal the layout of the farm in an interesting and satisfying way. Find good start and end frames for your panning sequence, and some motivation for it, so that it makes a point and does not just appear an arbitrary camera movement. For instance, you could start on a picturesque subject, such as a quaint, thatched farmhouse and pan round to end on a modern milking parlour.

Always rehearse panning and tilting shots two or three times before actually shooting a sequence, so that your camera moves are smooth and steady, and you will know exactly

▲◀ Pigs are natural video stars – the muddier the better. Steady your camcorder on the fence and zoom out from a close-up.

where to stop to achieve a well-composed final frame.

By now the kids will be getting impatient to see the animals. Some of these may be housed inside gloomy sheds where there simply won't be enough light to get good results. In other cases, the layout of the pens may be such that you can get good shots of the animals, but only the backs of heads of the visitors. Do not despair! Just take off your video-maker's hat, relax and enjoy what is on offer – a few 'natural breaks' in the shooting will ensure that making the video doesn't dominate the day.

When you come to a pen that is well-lit and accessible from more than one angle, you can position yourself at 45 degrees to those looking-on and get very good shots of the interaction between the children and the livestock!

▲The mid-shot of patting the horse serves to introduce the point-of-view shots that follow.

ANGLES

You can vary the height of the camera to bring out the contrast between the size of the children and the various animals they encounter. For example, if you are taking a shot of a horse looking over a fence, shoot from a low angle, reflecting the child's point of view. Then go for a reverse shot, showing the horse's-eye view of the child, from a high angle. You can also reinforce the impression of mutual inspection by making use of complementary eyelines. Thus if the child is looking up and towards the left of the frame, the horse should be videoed from a position where it is looking down and to the right of the frame.

Be prepared to let an entertaining sequence such as this run its course. Frame up so that the shot is wide enough to see all the action, but close enough to capture your child's eager expression. The magic is in the changing reactions of the child, ranging from initial hesitancy, through growing confidence to complete enchantment as the creature allows itself to be stroked, or gently nibbles some hay held out to it.

The child's-eye view

Try to get down for a low-angle shot at the child's eye-level, resting the camera on the cross-strut of the pen, if possible. Beware of getting out-of-focus pen bars in the foreground of the shot, as they will come out as a distracting blur.

Resist the temptation to shoot everything that moves, or you'll end up with hours of material. Ducks, geese and cockerels, parading round the farmyard with their characteristic waddle, glide or strut, can make amusing bridge shots between one sequence and another, but keep

▶ **Most farm animals are friendly and will need little encouragement to come towards the camcorder – especially if they think food might be on offer.**

OFF-SCREEN SOUND

Sound coming from a source outside the frame can lend a vivid three-dimensional quality to your shots, suggesting a whole world existing beyond the frame of your picture. For example, while you are framing a cage of rabbits the dominant sound may be the farmyard cockerel. If this is a problem, one way of dealing with it is to pan round to reveal the source of the sound.

▲ **As a goat reaches the fence to munch a snack, move round slightly to get a shot capturing your child's reactions.**

them short. After panning with the bird for a few seconds, leaving room in the frame for it to move into, ease the video camera to a standstill and allow the c hicken to move out of shot. Hold the empty frame for a beat. This will give you a steady cutting point to your next shot.

Contrasting scenes

The farmyard is a place of wonderful natural contrasts, so try to be aware of any opportunities to vary the mood of your video. Thus a shot of a pair of obese, slumbering Vietnamese pot-bellied pigs will have the maximum impact if it is placed for contrast between the frenzied activity of a goose bathing himself in a muddy puddle, and the chocolate-box prettiness of a litter of kittens playing around in a hay bale.

When you have got the best out of the animal talent, it is time to move on to the machinery, and preferably machinery in action. If you can't be around at milking time, see if the dairymen will let the kids pretend to be cows! They will have great fun lowering their heads to activate the automatic feeding troughs and trying out the milking clusters on their thumbs, and you may end up with a hilarious scene.

Then make your way to any fields where there is hay-baling, combine-harvesting or ploughing going on. These tasks are methodical, repetitive and relatively slow, giving you ample time to observe the pattern of activity and decide on shooting strategy. Pick out the really interesting

◀ **Most farms have old tractors that make ideal climbing frames for kids to pose on before it's time to go home . . .**

visual moments, and make a mental note of any long-drawn-out and less exciting sections you can safely leave out.

By alternating shots of the farming processes with brief cutaways of the family watching, you need only include the best moments, thus keeping your sequence reasonably compact. If it seems safe, and won't involve trampling any crops, you could position yourself in front, and slightly to one side, of the advancing tractor, and walk backwards (you will need to get another person to guide you!). When you want to end the shot, stand still and let the machine lumber out of the frame. Change your camera position to a side angle to shoot some close-ups – the massive tyres trundling over bumpy ground, or perhaps the weather-beaten faces of the farmhands at work.

If it is safe to do so, and providing it doesn't interfere with the farm's work, follow the machines while they are in operation. For example, you could show a hay-baler as it ingests g loose mounds of hay and disgorges them as processed bales. Try to include a wide shot every now and then. This will serve to show your video viewers how the field is slowly worked over by all the machines, and will reveal the change between the fields that are already harvested and those with crops that are waiting their turn.

Here as in any video, what you will want to obtain are a few telling shots that will link into a narrative, enabling the viewer to follow the film easily. Together with well-chosen and colourful detail shots, these will provide some memorable moments for your video archive.

Q ***When videoing my daughter's rabbit in its cage, I found it hard to get good, in-focus shots of the rabbit inside. Where am I going wrong?***

A *The wire mesh in the foreground has 'confused' the Autofocus. Switch to Manual, zoom right into the subject to focus up, and then widen the shot to your chosen area.*

OUT-TAKES !

FUNNY FARM
Dark sheds and curious animals may cause a few problems

• **The hungry goat** – Getting this close could result in an amusing shot – or just a well-chewed lens cap!

• **Dark corners** – Some places on a farm, such as this pig-sty, are too dark for good shooting without a lamp

• **Remotely interesting** – This machinery would have much more impact if it were viewed from closer in

HAND-HELD MICROPHONES

NO REALLY SERIOUS VIDEO BUFF WOULD BE WITHOUT A COUPLE OF ADDITIONAL MICROPHONES. WE LOOK AT WHAT THEY DO AND HOW THEY IMPROVE YOUR RECORDINGS

The microphone built in to the camcorder is perfectly adequate for most general situations but, sooner or later, the time will come when you will need an alternative, perhaps more specialist type of microphone. Once such type is known as a hand-held microphone because it is not fixed on to the camcorder but is attached by a plug and lead and held towards your subjects in the hand.

▼ Hand-held mics, powered by small internal batteries, can supplement the sound pick-up of the camcorder's built-in mic.

POLAR RESPONSE

All hand-held microphones have a particular range of sound pick-up, and this is known as the Polar Response. Microphones which can pick up the sounds to the front, the sides and the back of the camcorder are generally known as Omni directional microphones, those that pick up sounds to the front and sides are Cardioid or Uni-directional, and microphones which pick up sound from a very narrow range from the front

BELOW: BRUCE COLEMAN

▶ **Omni-directional mics pick up sounds from all around and are suitable for occasions such as children's parties.**

only are known as Super-cardioid and Hyper-cardioid.

A microphone's Polar Response is the first point to consider when looking at microphones; Omni-directional mics are useful when you have several people to record in a group (children's parties, wedding receptions, etc), but your camcorder's built-in microphone will probably be Omni-directional and be suitable for these occasions.

Cardioid microphones are more directional, and are useful for those occasions when you want to record the sound from a particular subject and cut down on some of the background noise. Singing is a very good example of this.

Super- and Hyper-cardioid are specialist microphones, and these enable you to record sound on wildlife videos, or voices at a distance clearly.

BUBBLES

◀ **Large Hyper-cardioid mics pick up the sound from a very narrow field in front. They are particularly good for recording wildlife soundtracks.**

▼ **Stereo cardioid mics give stereo sound – ideal for moving subjects.**

FREQUENCY RESPONSE

The next specification you need to look out for is called Frequency Response. This is a measure of what range of sounds can be picked up by

THE NATIONAL MOTOR MUSEUM, BEAULIEU

L—STEREO—R

the microphone. You will see it written in terms such as 100Hz to 12KHz. (The Hz is Hertz, the unit of frequency. KHz is Kilo Hertz, one thousand Hertz.)

Low numbers (20Hz to a few hundred Hz) are bass frequencies. High numbers (8KHz to 18KHz) are treble frequencies. Those in the middle are mid-frequencies.

Out of range

A microphone cannot necessarily respond to all the frequencies you can hear. A typical reponse would be 100Hz to 12KHz; remember that the camcorder microphone can only record a certain range of frequencies, perhaps 80Hz-8KHz on normal and 20Hz-20KHz in Hi-Fi. So what you hear is not always what you get!

There is no point spending lots of money on a microphone that can pick up sounds that a camcorder cannot record, so before you buy, do check your camcorder's specifications. Remember that traffic rumble is a very low bass sound, so if you want to avoid this noise, go for a mic with a starting figure of around 2000Hz. There is very little useable sound over 12KHz (except music), and your ears cannot hear anything over 16KHz to 18KHz.

Camera Sense

HOT SHOE SOUND

A useful alternative to a hand-held microphone, this battery powered hot shoe-mounted super-cardioid mic makes a great supplement to the camcorder's own omni-directional microphone.

Such highly directional pick-up is particularly good if you have problems with operation noise (the zoom lens operation and Autofocus sound coming through on your recordings), and the integrated windsock effectively cuts down on wind noise. The hot shoe mounting also leaves both hands free to deal with other important tasks such as manual focusing and steadying the machine.

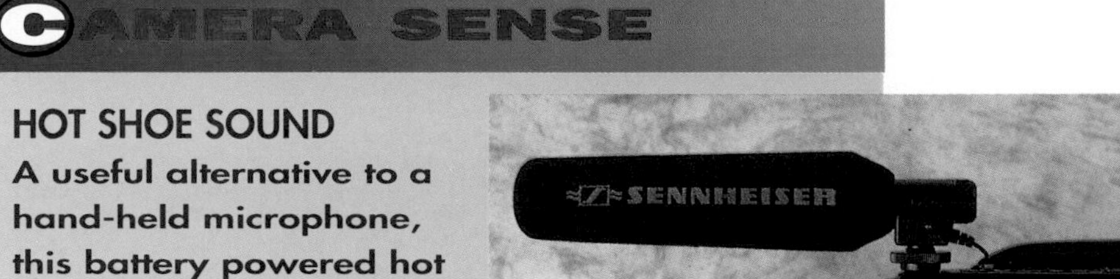

ALL MICROPHONES COURTESY OF SENNHEISER

REX FEATURES

◀ Cardioid mics have a heart-shaped pick-up which cuts down on unwanted sound around a subject.

▶ The polar response of mics varies from the all-round pick-up of the omni-directional type to the directional response of a super-cardioid model.

POLAR RESPONSE

Omni-directional

Cardioid

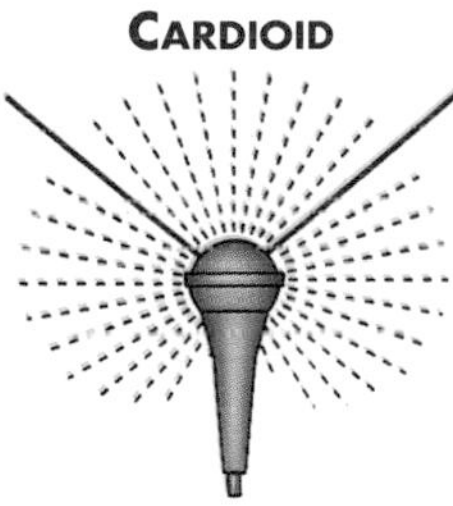

Stereo cardioid

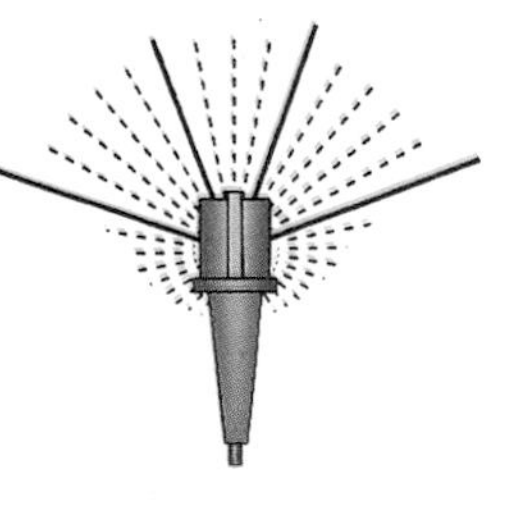

Super-cardioid

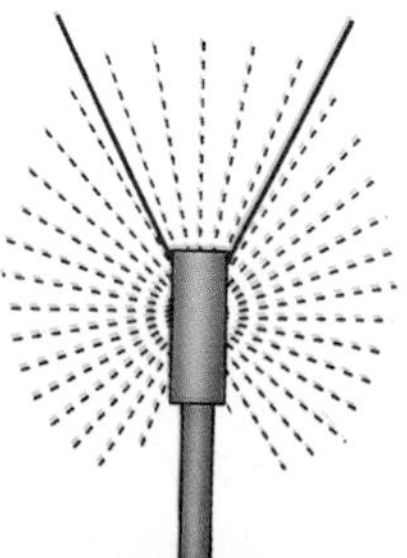

Microphone types

The third specification relates to the way the microphone works. The two main types are called Dynamic or Moving Coil, and Electret or

CHECK IT OUT

Before you buy a hand-held mic:

- Decide which kind of polar response would be most useful. Cardioid is a good all-round starter, but for specific sounds choose a super-cardioid.
- Think about the frequency respons you need. 100Hz to 12KHz is good enough for everything but live music.
- Ask about compatibility with your camcorder. If you take it along with you, the dealers may let you try the microphone before you buy.
- Check that the plug fits your machine – if not, you can buy an adaptor.
- If you think you need a long cable, buy an extension lead. This is easier than trailing a fixed long cable everywhere with you if you don't need it all the time.

▲ If you can't decide which mic you need, or feel you could use all types, boxed modular systems, offering a choice of pick-up, are the answer.

Capacitor. Those of the Dynamic family are good, dependable all-rounders; you can carry them just about anywhere with little risk of damage. The Capacitor family are more sensitive; they dislike damp conditions and can be temperamental if they get cold. Although they generally have a better frequency response and can pick up quieter sounds than Dynamics, they need a battery to make them work.

LISTEN CAREFULLY

Wherever possible, try to listen to the actual sounds you are recording through the microphone. Connect a pair of headphones (the closed-back, ear-muff type work best) to the headphone socket of the camcorder. This enables you to check whether the mic is working to your satisfaction and also to monitor exactly what it is picking up. But remember, recording sound without listening to it through a set of headphones is like recording pictures without looking through the camcorder's viewfinder! □

CRADLE ATTACHMENT A useful accessory which enables you to fix the mic to a variety of supports – especialy handy when interviewing. Some are sold with the mic, others are sold separately.

PLUG AND LEAD Make sure the plug fits your camcorder! If not, buy an adaptor. The thicker the lead, the better the sound, and the less liable it is to damage.

CAPSULE COVER This plastic or wire mesh protects the microphone. Covers vary in size, and can comprise built-in windshields. If not, they can be bought separately.

MIC CAPSULE 'The microphone' itself. This section is sensitive to shock and damp, so treat it gently.

BODY This is the bit you hold, so choose one that feels right for you. Metal is more durable than plastic, and matt-black or grey more practical, as there is less danger of reflections flashing in to your lens.

BATTERY COVER Electret mics are powered by small batteries which fit neatly into a housing within the body. Batteries are generally sold with the mic.

YOUR VIDEOS ON TV NEWS

GET INTO THE HABIT OF TAKING YOUR CAMCORDER EVERYWHERE WITH YOU, AND YOU MIGHT JUST HAPPEN UPON AN INCIDENT THAT WOULD MAKE THE NEWS. SO WHAT DO YOU DO IF YOU SHOOT SOME VIDEO FOOTAGE BEFORE THE PROFESSIONALS ARRIVE ON THE SCENE?

A passing motorist sees a group of white police officers beating a black man, Rodney King, in a Los Angeles suburb. Coincidentally, he has a video camera with him and he shoots a cassette of the incident. Nearly a year later, when the officers are acquitted of assault charges, those amateur pictures spark off days of fierce rioting when they are broadcast on television. Within hours the video appears on news reports of the riots around the world.

A decade or two before, an incident like this, where an amateur both recorded, and by doing so then made the news, could never have happened. Heavy, difficult to operate and hugely expensive, very few people would habitually take a video camera out with them in the 1970's. Yet even today, with the boom in lightweight, sophisticated camcorders, the chances of someone just happening on a top news story are still small.

Nevertheless, it does happen. On Rememberence Sunday 1987, both the BBC and ITN used amateur footage shot by a spectator of the

▼ In July 1988, a tourist in Greece was able to get shots of a cruise ship after it had been stormed by terrorists.

PHOTOS BY REX FEATURES

RATION DAVID ASHFORD

brutal terrorist bomb attack in Enniskillen which left eleven dead. Four months later, a tourist video of four IRA members shot dead in their car in Gibraltar, fuelled a massive controversy about whether the British army had operated a so-called 'shoot to kill' policy. Amateur shots of the M1 plane crash, the Soviet cruiseliner Maxim Gorky after it struck an ice flow, and the devastation in a London pub after an IRA bomb, all appeared on national news bulletins.

Should you have shot something you think is newsworthy, your first step is as obvious as getting out the phone book and finding the number of the newsdesk of a network TV station. Nowadays, a standardized format and sophisticated technology mean thel quality of amateur video is fairly high. However, as technical quality falls dramatically when you transfer video even just once, they only accept original tapes – which they return later. These can then be transferred to broadcast standard tape for editing and transmission.

Getting your pictures to the newsdesk is the next step. Depending on the urgency of your story, the network can liaise with a local television station to arrange for your pictures to be sent via a land line, or, if you are abroad, a satellite link. Otherwise it is a matter of express post or courier – all of which the TV station would pay for, whether the tape is used or not.

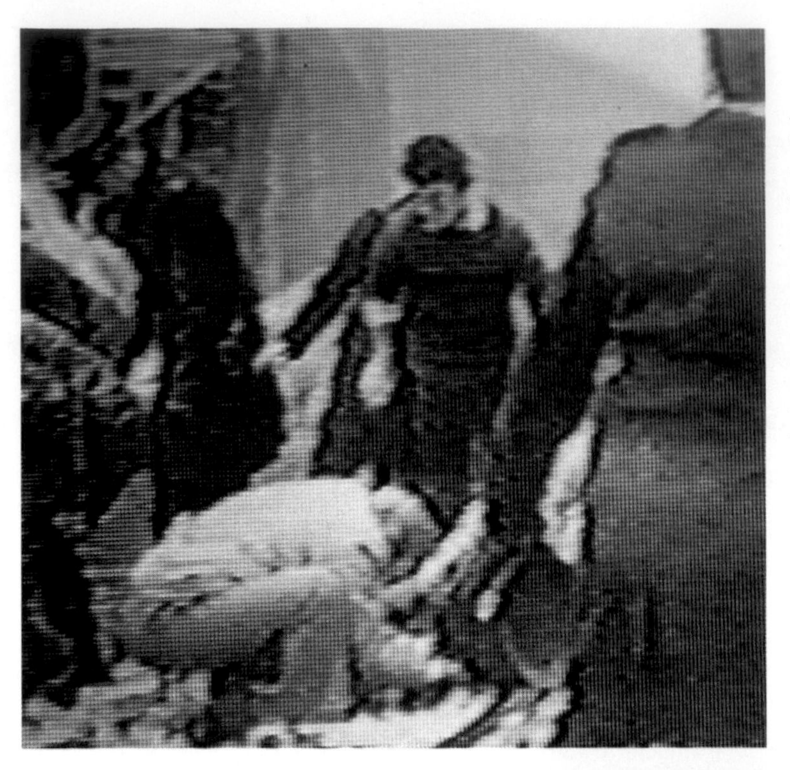

◀ Enniskillen, 1987. The horrific scenes following the terrorist bomb attack on Rememberence Sunday in Northern Ireland were captured on amateur video and broadcast to the world on network television.

Q

Would getting my video on the news help me to find a job in TV?

Unfortunately, no there is a very high rate of unemployment even among trained camera operators with many years of experience. But having shown you have got a good idea of what stories make the news will surely help when applying for a news job.

A

▶ Some of the first broadcast footage of the British Midland plane crash near Leicester in 1989 was supplied by an amateur.

While all news programmes say they are prepared to pay for material used, they stress there are no fixed rates for amateur video. And they are adamant they would be unlikely to run candid shots of a celebrity or Royal if money becomes an issue.

The majority of amateur videos which have appeared on British television news have been used for a nominal fee, or even for nothing. According to the editors who have accepted them, people have just wanted to help in the wake of a disaster or accident.

For several years now, Gatwick Airport viewing tower has had a daily contingent of amateur video enthusiasts, two-way radios tuned in to air traffic control, just waiting to capture a major air disaster. This may seem an overly-detached approach to newsgathering; but as camera prices continue to tumble, and the technology to soar, it will become increasingly common for amateur buffs to pip the professionals to the post. □

Steady as you go!

PHOTOGRAPHS BY LYNDON PARKER

USED WITH CARE AND IMAGINATION, THE HAND-HELD VIDEO CAMERA CAN LEND A SENSE OF ENERGY AND EXCITEMENT TO YOUR VIDEOTAPES

Today's camcorders are designed for hand-held shooting. Their lightweight and compact design makes them easy to carry and flexible enough to be used in almost any situation. However, shooting by hand has one great drawback that will be familiar to all videomakers – that of camera shake. Even camcorders with integral anti-shake devices can be affected by the movements of your hands and body as you shoot – movements that will be faithfully reproduced on the screen. Even if the viewer does not immediately notice, the cumulative effect of viewing a recording with continuous camera movement is disconcerting, not unlike the effect of motion sickness. But there are a few

- ***Shoot without excessive camera shake***
- ***Pan and tilt while filming hand-held***
- ***Shoot while moving with the action***

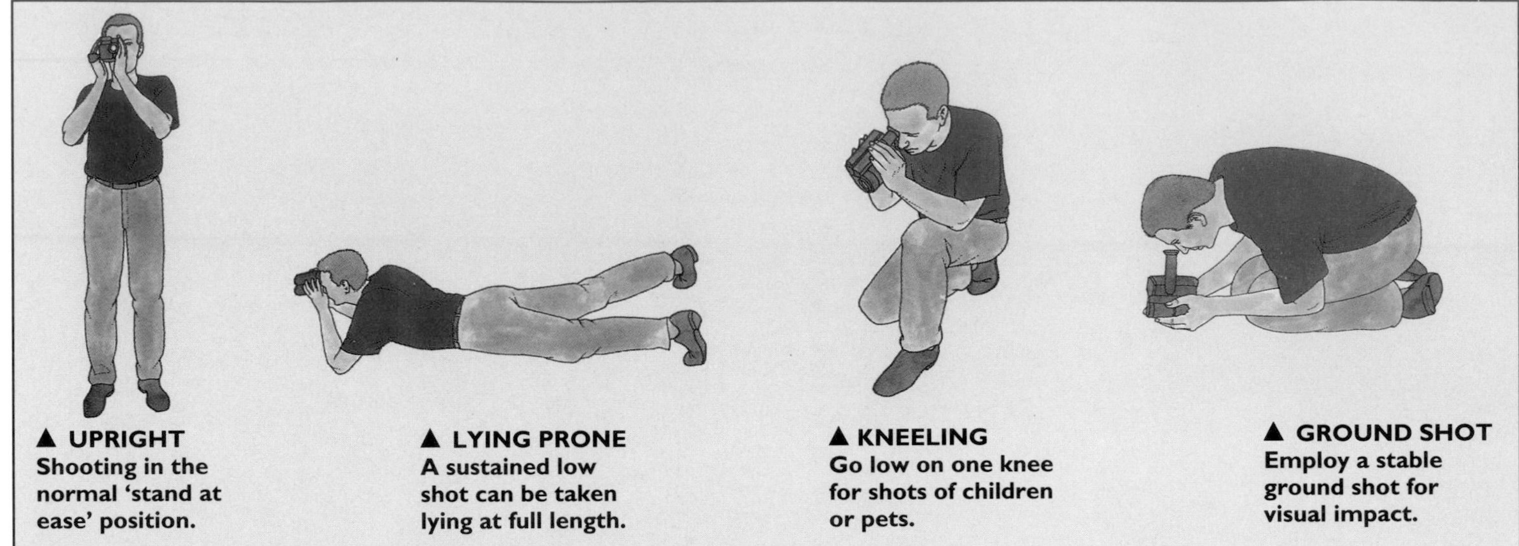

ILLUSTRATIONS BY ROB SHONE

▲ **UPRIGHT**
Shooting in the normal 'stand at ease' position.

▲ **LYING PRONE**
A sustained low shot can be taken lying at full length.

▲ **KNEELING**
Go low on one knee for shots of children or pets.

▲ **GROUND SHOT**
Employ a stable ground shot for visual impact.

simple rules you can follow to ensure your camera work is as steady as possible and to minimize shake.

The first, and most important, rule is to use both hands to grip the camera. Unfortunately, a lot of camcorder advertisements try to emphasize the simplicity of operation and ease of use of their products by depicting the videomaker using only the right hand. With automated functions, this is perfectly possible, but it is not generally good practice. Remember, your left hand is free and you can always use it to steady the camcorder and operate the controls.

Adapting your grip

The method of supporting the camcorder will differ from model to model and will be influenced by how much manual control you want.

With the smallest 'palmcorder' models, tuck the heel of your left hand under its body, with your fingers pressed against the left side. Your hands will be cupping the camcorder between them.

On the conventional mid-size models, the zoom lens and its housing extend farther forward – as does the centre of gravity. So support the instrument with the fingers of your left hand under the lens barrel and your palm against the body. In most cases your fingers will then be close

CAMERA SENSE

SHOOTING AT A LOWER LEVEL

Sometimes you will want to get low-level moving shots – following a pet, for instance, or a child at floor level. You can improvise a camera cradle with a strong plastic bag. Place the camcorder at the bottom of the bag and cut a hole for the lens, making sure that the autofocus window is not obscured. By dangling the bag close to the ground you can shoot while walking along. Set the zoom lens to wide-angle, switch to record, then put in the bag. Aim carefully, since you will be shooting 'blind'. This set-up can also be an effective camouflage for the camcorder when you want to sneak candid shots of family or friends.

PHOTOGRAPH BY INGO JEZIERSKI / THE IMAGE BANK

▲▶ **ANGLED UP**
Rest the camera on the ground, tilted up for added drama.

▲ **OVERHEAD**
A fully rotatable viewfinder enables you to lift the camera to shoot over the heads of the crowd. If you remove the viewing lens, you can use viewfinder in sportsfinder mode.

to the focus and zoom controls.

For models with heavier interchangeable lenses, support the lens itself with one hand while operating the controls with the other hand.

Full-size machines are designed to rest on the shoulder and your cheek can be pressed against the body of the camera to provide additional stability. Nevertheless you should still use two hands. In most cases your left hand should be gripping the focus ring, both for support and for manual focus control.

Your grip should always be firm but not tense, to avoid nervous shake, which can easily be transmitted to the picture. Your arms should be tucked in firmly against your chest.

A BALANCED POSTURE

The second rule of avoiding camera shake is to ensure that you are in a balanced and secure position. For maximum and natural balance the ideal standing posture is something like the military 'stand at ease' – feet firmly planted about 30cm apart, toes pointed slightly outwards, and legs straight with rigid knees.

With hand-held shooting there is a temptation to keep the camcorder at shoulder level. But you should shoot at different heights to vary your angles and add visual variety to your coverage of events. By going down on one knee you can support your elbow with the other and provide a steady position for long takes. This height is particularly good when recording young children or people sitting in low chairs.

You can also kneel with the camcorder cradled in your lap. For this position you will not have your eye directly to the eyepiece so you may need to use the 'sportsfinder', removing the viewfinder lens and cup.

For shooting at the lowest angles, lie full-length on the ground, propping yourself on your elbows. A cushion may be useful here for extra comfort and support. This is an excellent position if you are trying to record a baby or toddler crawling or walking towards you at ground level.

For a very low angle and a steady

SOUND ADVICE

TREAD CAREFULLY

Moving while you are shooting increases the risk of inadvertently recording handling noises. Keep a firm grip on the machine and do not let loose clothing brush against the body of the camcorder. Swift camera movements will create a wind-like noise on the microphone, which is another reason for keeping your pans and tilts slow – as they should be anyway if the viewer is to take in the scene.

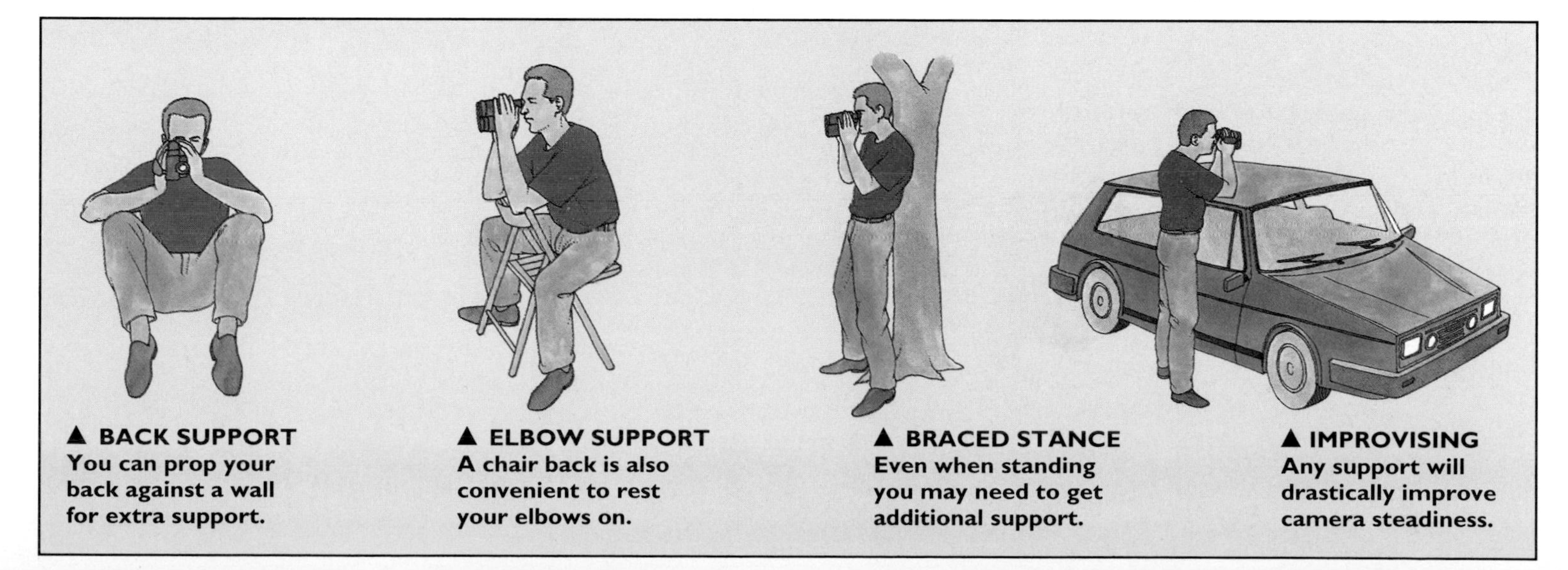

▲ **BACK SUPPORT**
You can prop your back against a wall for extra support.

▲ **ELBOW SUPPORT**
A chair back is also convenient to rest your elbows on.

▲ **BRACED STANCE**
Even when standing you may need to get additional support.

▲ **IMPROVISING**
Any support will drastically improve camera steadiness.

DANGERS WHEN MOVING

Looking through a viewfinder while moving is hazardous. When you move sideways or backwards, a helper's hand on your shoulder will guide you. Even when moving forward, keep both eyes open. What you see with your non-viewfinder eye will give a truer sense of your position.

shot, kneel and rest the rear end of the camcorder on the floor, tucking your fingers under the front to raise it to the appropriate angle. Use this for dramatic shots of street scenes or striking views of pets.

Some viewfinders rotate downwards by 90 degrees. You can hold this type of camcorder above your head and shoot over the top of a crowd. Adopt the normal stance, with your elbows slightly cocked. Again, it can be helpful to remove the eyepiece and view the monitor screen direct. However, this can be a tiring position and you might find it hard to avoid camera shake.

SEEKING SUPPORT

The third rule for steady camera work is always to seek additional support when you can. Indoors, you can lean against a wall or doorway. You can sit in a chair and use the arms to support your own, while the camcorder is cradled in your hands. Use a table or the back of a chair to support your elbows, or sit on the floor with your back pressed up against the wall and rest your elbows on your knees. Outdoors, you can use trees, lampposts and walls to provide body support, and cars, low walls or fences for elbow support.

The fourth rule for avoiding cam-

IMPROVE YOUR SKILLS

At the Top of the Stairs...

An exciting but difficult use of the hand-held camera is the suspense-filled ascent of a staircase to a revelation in an upstairs room.

▲◀ The way ahead is disclosed by an upward pan from the foot of the stairs.

▼ 'Crab' around to the first step. Move slowly and deliberately, with knees slightly flexed.

▲▶ As the climb begins the camera angle is raised to show the whole flight of stairs. It is vital to avoid a jogging motion here.

▲▶ **A moment of tension builds as the camera approaches the top of the stairs. Do not vary the pace as the landing beyond is revealed to be empty.**

◀ **The tension mounts again as the inexorable advance continues, now towards the door.**

▶ **A tilt down reveals a shaft of light thrown on to the floor from one door. Skillful use of lighting in this way can help create an illusion of mystery and suspense.**

◀ **For a smooth final advance into the room, you will need a helper out of shot to open the door.**

era shake in hand-held work is to shoot with as wide an angle as possible and a supplementary wide-angle converter lens, which you can buy from all good video stores, is often a valuable asset. Try to restrict your use of the telephoto end of the zoom when shooting hand-held – just as this magnifies the image, it will also magnify the shake.

Hand-held shooting will also involve camera movements such as pans, tilts and cranes. For hand-held pans, the secret of success is to remember to keep your feet facing in the direction you want the shot to finish. Turn the camcorder to where you want the shot to start by twisting at the waist. Once you start shooting

Q

What is the best way to record steady, hand-held shots of horse races and horse-jumping?

Try setting your camcorder to a faster shutter speed. As with still photos, this can improve image definition when you are shooting swiftly moving subjects hand-held.

you simply unwind until you are facing the finishing position. Working the other way round will result in jerkiness or movement of your feet at the end of the pan. The 'untwist' method ensures that you don't try to pan too far. The same technique can be used when kneeling on one leg.

With the camcorder held at shoulder level, a hand-held tilt involves leaning backwards, arching your back as you move. This technique will also work while you kneel on one leg. Alternatively, cradle the camcorder at waist level and tilt it gently towards your chest.

You can also execute successful hand-held 'crane' shots, in which the camera rises to follow the action or

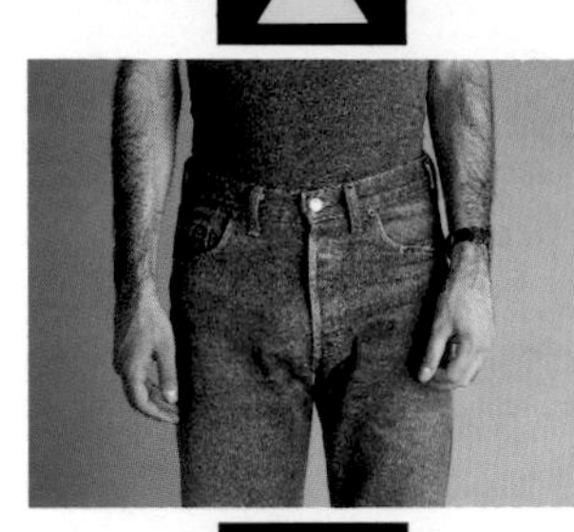

▲ **'CRANE' UP**
Rise up from one knee to a standing position.

reveal a subject. Start in a stable position on one knee and simply push upwards by straightening your other leg until you are fully upright.

KEEPING FOCUS

Remember that when you move with the camcorder you are likely to be changing the camera-to-subject distance. Autofocus systems should cope with this when there is not too much movement within the frame. But if you are relying on manual focus, you may have to make adjustments as you shoot.

Moving with the camera calls for a particular way of walking. Crouch slightly, this will force you to make shorter strides, with less rise and fall than normal. Make each footfall as soft as possible and aim for a gliding, slow-motion movement. Apply the same principles when you are walking backwards with the camera. To 'crab' – move sideways – keep the same posture but swing one leg in front of the other as you move, treading with deliberation.

In the end, however good you become at hand-held shooting, it is best not to rely on this approach alone. Using a tripod not only provides stability but enables you to make full use of the telephoto. It enforces a natural discipline in the way you shoot and extends the range of subject matter you can successfully cover.

You may have noticed how television professionals intersperse hand-held sequences with other footage, such as an interview, that has been shot using a tripod. Some news stories may well be shot entirely hand-held, and fly-on-the-wall documentaries use the technique for intimacy and impact, but it is rare for viewers to find themselves watching hand-held material for more than a few minutes. Bear this in mind and try to use a tripod when possible – but don't lose the spontaneity and freedom of hand-held shooting.

SHOOT IT STRAIGHT!
With hand-held shots, keep horizontal and vertical lines correctly aligned.

PHOTOGRAPH BY DAVID MAENZA / THE IMAGE BANK

• **LEANING TOWERS** – BUILDINGS CALL FOR PARTICULAR CARE. CHECK VERTICALS AND HORIZONTALS AGAINST THE EDGES OF THE FRAME

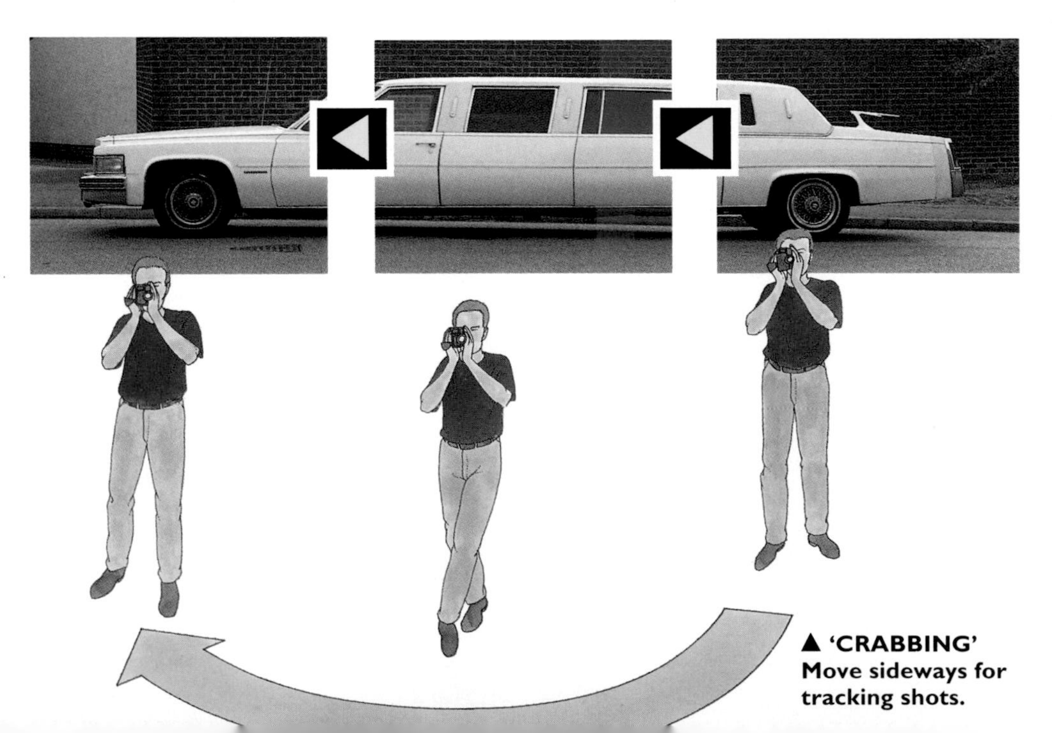

▲ **'CRABBING'**
Move sideways for tracking shots.

Editing for creative impact

THE ORDER IN WHICH YOU PLACE THE SEQUENCES OF YOUR EDITED VIDEO MAKES ALL THE DIFFERENCE TO YOUR END RESULT, SO PUT YOUR ASSEMBLE EDITING SKILLS INTO PRACTICE AND START EXPERIMENTING!

Every video should tell a story, and every story should have a beginning, a middle, and an end to give it a satisfying shape. This shaping begins with decisions about what you actually put in front of the camera at the time of shooting, but it really comes to the fore at the editing stage.

With editing, you can rearrange time and space. It can be one of the most satisfying stages of making a video, allowing you to create a rhythmic flow in your programme by developing a variety of pace, a contrast of scenes, a balance of different elements.

CHANGING THE MEANING

Early in the history of the moving image, it was discovered that the meaning of a shot is not absolute, but may depend on what image is placed next to it. Experiments were carried out to show that a close-up of an actor with a neutral expression appeared to the audience to take on a specific meaning according to which shots it was intercut with: thus the actor's face appeared to be expressing happiness, distress or desire, according to whether it was intercut with shots of children playing, a coffin, or a beautiful girl. In the same way, the effectiveness of your own home videos can be enhanced by the order in which the shots and sequences run.

ILLUSTRATIONS BY NICK PAYNE / THE GARDEN STUDIO

▲ You should already have reviewed part or all of your original tape (see Practical Editing, pages 6 and 7) in preparation for editing, made an edit log, and decided which sequences you want to cut to compress the action. Now jot down a plan for a complete edited version of your day on the beach.

CAMERA SENSE

MAKING A MONTAGE

A montage is an assembly of shots which are linked thematically and do not necessarily have to show continuity in time or space. If you want to create a montage of a child swimming, the same child might be seen successively: diving under a wave; vigorously splashing a brother or sister; cavorting in the foam as it breaks on the beach. Montages are great fun to create, and they are often cut to a piece of music to retain sound continuity.

Once you have assemble edited your first few simple videos, you can start experimenting with the order of your material. After all, what you are trying to do is capture the spirit of an occasion, rather than reproduce a faithful catalogue of events.

BACK TO THE EDIT LOG

Whenever you assemble edit, it makes sense to make an edit log first. However, if you intend to re-order material, the edit log becomes essential, otherwise the energy which should go into creative editing will be dissipated in frustrating searches back and forth along the tape to find the material you need to recreate your new version.

How much you manipulate the material is very much up to your individual taste and intentions. You may have a purist desire to 'tell it like it was' with very little intervention,

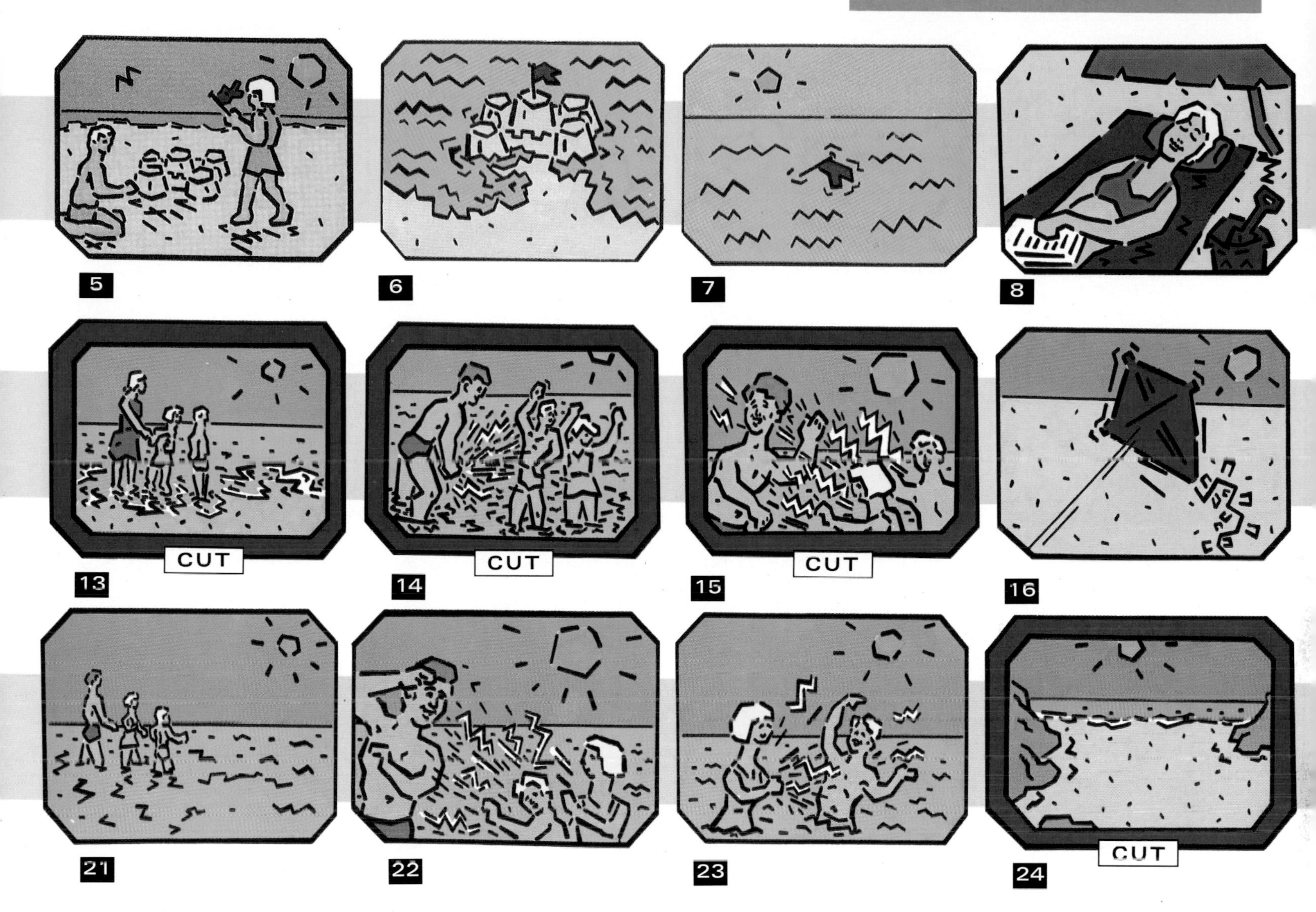

you may have a humourous streak and want to reorganize the material for comic effect, or a romantic nature which inclines you to select the more moody atmospheric shots.

Editing is often thought of as a 'musical' process, and it is sometimes helpful, when thinking of overall structure, to think in approximately musical terms. Thus, reviewing your material of a day at the beach, you might begin to sketch out a rough framework: finding a spot to sit – lively up-tempo; sunbathing – leisurely, ballad pace; beach cricket – brisk choppy rhythm.

A list of the new order you want events to run in is important at this stage, and this running order is dependent upon the editing technique you choose. Make a rough plan of how the story will unfold, bearing in mind the pace and flow of each individual sequence, as these may affect your editing style.

Some scenes may be shot in a particular order for convenience, but need to be intercut to make visual sense. When shooting the beach cricket game, for example, you may have decided to concentrate first on the batsman receiving a series of balls, then on the bowler delivering several balls, then on the fielders, getting long steady shots of each, rather than following the action frantically.

A NEW ORDER

At the editing stage you will obviously want to re-order the material, choosing the best moments from each set-up to reconsruct the game. Such a series might be ordered as follows: 1, batsman prepares to receive; 2, fielders take up position; 3, bowler starts to run; 4, cut back to batsman waiting; 5, quick cuts in succession – the bowler bowls, the batsman strikes, the fielder reaches to catch.

▲ Assemble edit your chosen sequences in the order you require, following the instructions on Practical Editing, pages 14 and 15. Always return to your original tape when making a new edited version. If you copy previously edited tape, the loss of sound and picture quality will be substantial.

FINAL EDITED VERSION

Establishing shot of beach

Applying the suntan cream

Into the sea with Dad!

Mum relaxes in the sun

Splashing out

Getting thoroughly wet

Ice creams for all

Delicious!

Mine's good too!

Making a sandcastle

Finishing touches

Let's fly the kite

It's taking off!

High in the sky

The evening tide approaches

Goodbye castle, goodbye beach

▲ Your final, condensed video of your day out. The superfluous sequences have been cut and the order has been completely re-arranged to create a new beach story with a beginning, middle and end.

Alternatively, you may have material of two activities which were going on at the same time – the older children exploring rock pools, and the younger ones building a sandcastle. You can create a visual effect of simultaneous or parallel action by cutting back and forth between the two scenes.

While these scenes are completely constructed out of cuts, you might try a different, more languid technique with the picnic lunch, lingering on an opening shot of the food, then cutting to expectant, hungry faces, followed by individual members of the family tucking in to sandwiches, salad and fruit. Use the same technique later with ice-cream.

Once you start thinking in terms of restructuring your material, all sorts of imaginative possibilities present themselves, and you may be pleasantly surprised to discover what a distinctive and individual flavour you can give to quite ordinary video material, with a bit of creative planning and editing. □

The sky's the limit!

AN AIR SHOW IS A CELEBRATION OF MAN'S CONQUEST OF THE SKIES. THE CAMCORDER USER IS GIVEN UNRIVALLED OPPORTUNITIES TO VIDEO EXTRAORDINARY AIRCRAFT, HIGH-SPEED ACTION AND SPECTACULAR AERIAL STUNTS AND DISPLAYS

Imagine you are at a show at Biggin Hill or Boscombe Down. Suddenly there's an ear-splitting roar from the far left of the airfield, and you pick out the Red Arrows aerobatics team racing across the sky. At 600mph, they rocket upwards in front of you, perform a huge loop in the heavens in perfect formation, then dive, shooting off to your right, before splitting off to form a spearhead formation. Awe at this display of death-defying aerobatic skill registers on the faces in the crowd. But how on earth can this event be captured on video?

Air shows are, of course, exciting and atmospheric events. New aircraft are seen for the first time and highly trained pilots enjoy showing off skills and machines. It is a rare chance to see legendary aircraft in action, such as the Spitfire, the Hurricane, the Me 109, the Lancaster and the Vulcan. The entire event is choreographed for the spectator.

Obviously a display by the Red Arrows or some other aerobatics

◀ **Zooming out from a detail to reveal the whole. This shot gives the viewer a sense of discovery as the RAF roundel is shown to belong to a Lancaster.**

PHOTOGRAPHS BY ALISTAIR GRANT

team would make a scorching climax to your video. But in order to give your video some sense of the whole event and to build up gradually to this climax, you will need to film other aircraft in the air and on the ground, and show the crowd, or people you know, enjoying the show.

Plan in advance

Pre-planning can make the most of an air show as many events – almost too many – will be happening at the same time. You will need to be selective beforehand, and decide which events interest you most. Firstly, look at the programme: this will tell you the order of most events, and the times at which they take place. A map of the show, if one is obtainable, will help you to consider the best shooting positions. This will be especially useful when you shoot high-action aerial displays. Inevitably, with so many aircraft taking off and landing, there will be restrictions put on your movements: if you are able to leave the spectators' area, keep an eye open for taxiing aircraft! If no map is provided, perhaps you can make a rough one of your own using an Ordnance Survey map as a reference. There are rarely second chances in a tightly scheduled programme – especially as pilots and air-traffic control have to consider the safety of all aircraft, pilots and spectators when organizing a show.

Scene setting

After shots of your family or friends arriving at the show, and of excited crowds streaming in, it would make good visual sense for you to shoot an overview of the airfield itself, with all the aircraft on display. For this you'll need to get to some high vantage point: most air shows provide grandstands, so try and get to the back of one, and pan slowly across the crowd to the aircraft. Alternatively, there may be some room in the control tower. The overview is a

scene-setter; after this you can get down to details.

Aircraft come in all shapes and sizes: military aircraft, light planes, business jets, and vintage aircraft which have been lovingly restored by their owners. Vintage airliners and modern military transports are often parked around the perimeter or in public access hangars. When video-ing an individual aircraft, make sure you give the viewer a sense of the plane as a three-dimensional object with a specific function.

For example, you might come across a flying boat, such as the Catalina, once used by the military but now employed to fight forest-fires. They skim across a lake, filling special tanks at speed with some seven tons of water that they then dump on the blaze to prevent it spreading. Video the aircraft head on and then, still videoing, walk around the aircraft steadily in a 45-degree arc – this will give the viewer a full sense of the plane. Close up on the wing-floats and water tanks to show, in visual terms, that it is a specialist aircraft capable of landing on water and fighting fires.

▼ To follow a Vulcan bomber into the sky from take-off, pan steadily to ensure that the aircraft is held in centre frame all the way. Your camcorder's Autofocus should ensure depth of field and clarity of image.

SOUND ADVICE

INTERFERENCE

At an extremely noisy event like an air show, you are very likely to encounter problems with sound interference. The sudden clatter of a helicopter starting-up, an ear-splitting, supersonic bang, or an unexpected, low-level, high-speed pass with an afterburner thrown in, are sounds nicely calculated to destroy clarity of reception by the built-in mic! For the most exciting results, use external stereo-mics, if you can, to enhance the sound of on-screen action. You can also use a directional super-cardioid mic, which will help you to get rid of the chatter of a noisy crowd and isolate – at least to some extent – the sound of what you actually want to record on your video.

◀ **By adopting different positions on the airfield you can give variety to your video. In position 1, the pilots' enclosure, you may get 'human interest' shots of the flyers.**

A more interesting sequence might be one which gives a sense of the aircraft's history. For example, when confronted by an Avro Lancaster, focus on the Royal Air Force roundel, and the bombs drawn on the cockpit that indicated successful missions. At the editing stage, you may want to add your own commentary, identifying in your own words the type of aircraft and its history. Something like: 'The Avro Lancaster was the first long-range tactical bomber capable of carrying a full bomb-load into the industrial heart of Germany. Raids like

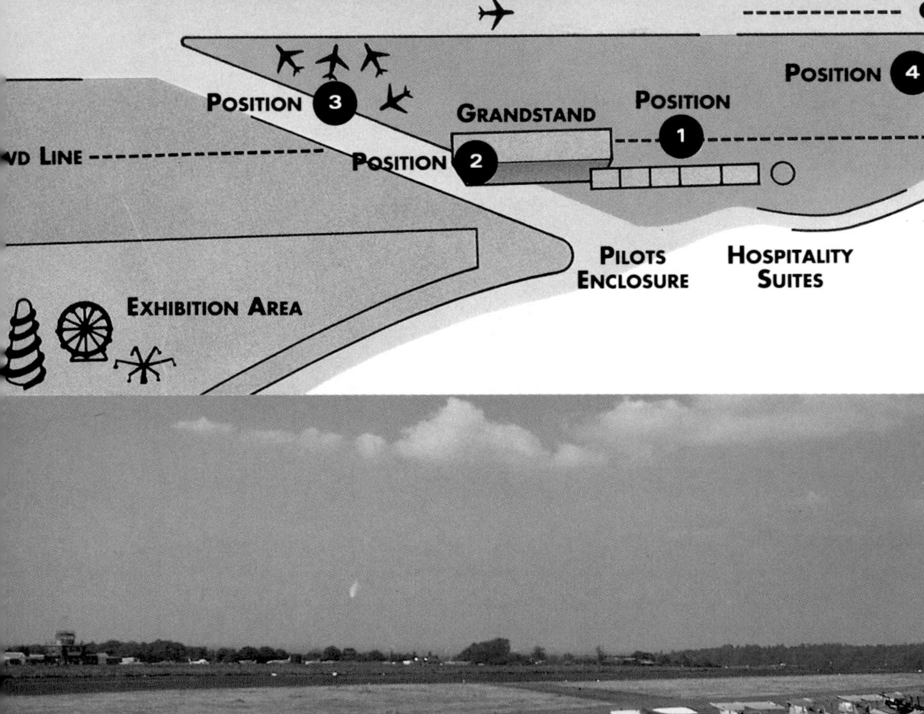

The Dambusters crippled Germany's war effort and helped win the war.'

World War 2 fighters feature prominently at air shows and there is nothing more stirring than a fly-past of Spitfires or Hurricanes. It is often possible to get quite close to these historic planes, even climbing into the cockpit. You can create imaginative links by cutting from a cockpit shot to a fly-past like that of the Battle of Britain Memorial Flight.

This brings you to the real thing – full-blooded action. There are a number of technical points you will need to bear in mind if you're going to shoot quality action footage.

◀ **To obtain good pans of the runway and the aircraft parked alongside, try and get into position 2, right at the back of the air show grandstand.**

▶ **No video of an air show would be complete without thrilling sequences of aerobatics.**

◀ **Close-ups of individual aircraft, such as this famous Lancaster, the *City of Lincoln*, can be obtained in the main aircraft exhibition and parking area at position 3.**

CAMERA SENSE

SHOULDER SUPPORT

Even if you have a large camcorder, you may choose not to use a tripod as this will limit your ability to move around freely. Check the camera sits comfortably on your shoulder, and stand with your legs evenly spaced so you can move the camera steadily by turning your body to follow any action. Select Manual focus, set the lens to infinity and you're away!

First, you should use a UV filter and backlight compensation, if you have this function, when shooting into the sun or at aircraft in the sky. Keep checking this so you do not lose crucial footage by forgetting to switch it on or off.

Secondly, for extra clarity, use telephoto to shoot fast-moving, distant small aircraft. Take a tripod with you to avoid camera-shake, and practise camera movements with it some days before the show.

CAUGHT ON THE WING

Most jets move at vision-blurring speed. To sharpen the definition of fast-moving aircraft, select a fast-shutter speed. If your camcorder has

▶ **To capture aerial action without obstruction from the viewing area next to the runway (position 4), you might find a stepladder useful to get head and shoulders above the crowd.**

CAMERA SENSE

COCKPIT CLOSE-UP

To convey some sense of an aircraft's specific role, crab in a 45-degree arc, starting from a head-on position, moving round and closing in on the cockpit – as in this sequence featuring a Jaguar Interceptor. The head-on position gives a good view of the inflight refuelling probe. By moving round to the final position, you reveal the Jaguar for the highly advanced aircraft that it is. The close-up shows the RAF roundel and the 'head-up' display behind the pilot, which feeds information to him via his computerized helmet.

▼ Move around the plane when shooting to reveal the power and hi-tech complexity of an aircraft from different viewpoints.

Head-on position

move around aircraft nose

zoom in for cockpit close-up

an inbuilt digital freeze-frame, you can use it to produce still-frame pictures of aircraft you particularly want to focus on. Record the control tower's public commentary as your sound-track on a separate cassette.

Don't overshoot; long takes of aircraft flying away from camera are boring. Fill the frame and shoot for peak action. Air shows are as much about aircraft enthusiasts as aircraft. Shoot atmospheric cutaways of people. Use these later to avoid long shots of aircraft coming round again for another low-level fly-past.

AEROBATICS DISPLAYS

Of the many events in an air show, the most eagerly awaited, of course, are aerobatics displays and formation flying. After all, you don't often see two planes hurtling straight at each other only to twist away at the last instant! To capture their marvellous manoeuvres on video tape, you will need to do a little research beforehand and to be on your toes on the day, as the action develops.

It is well worth trying to get hold of a Red Arrows performance brochure well before the show. This will probably contain a programme of the proposed display, with diagrams of various aerobatic manoeuvres. Choose the routines that interest you most, and plan shots to determine

▶ **Rather than ending up with a video of a lot of black dots against a blue sky, try and include some colour features like the blue and red smoke trails used during aerobatic displays.**

◀ **Cut away from the main action overhead and focus on the faces of spectators, intent on the display, for added interest.**

▲▼ **Look around you and pick out other details that are visually striking and that will add variety to your video.**

when you need to zoom in or pull wide. Rehearsing action through the camera with this plan will improve your final coverage of the event.

For example, the Arrows may propose to perform the Eagle Loop as part of their display. To shoot this, start to zoom as the formation heads into the loop; widen out as they enter the loop in order to capture its whole progression; close in as they come out of the loop and follow them out on zoom. In general, use telephoto for aircraft in close formation, and the wide angle for breaks. On the same principle, you'll need to use zoom as aircraft dive into The Parasol Break – the Red Arrows' spectacular finale – and then to pull wide quickly as they split. For a more difficult routine like the Shackle and Double Rolls-Synchro, where two aircraft cross over, fly in opposite directions, and one weaves back to cross under the path of other, you will have to content yourself with trying to follow the progress of one aircraft. On video, this will create a certain suspense, as the viewer will never be certain when the other aircraft will shoot in and out of the frame. Remember you don't need to record all the Red Arrows' routines to give the viewer a vivid idea of the skill of their display.

Audio involvement

To attain still greater viewer involvement in the performance, you can obtain a special aviation-frequency radio-receiver, and hold it up to the inbuilt microphone to record sound from air-traffic control and aircraft to communicate the tension, excitement and precision of the flying.

▲▼ **Children love air shows and this is an ideal opportunity for them to get close to real aircraft as well as see them on the wing.**

▼ **A shot of a Spitfire taking off into the wild blue yonder would make a spectacular final image.**

A team may change their routine if there is low cloud or bad weather. The control-tower will announce this but if you cannot hear it clearly, ask a steward and use your research to create a successful contingency plan.

CAPTURE THE ACTION

Lastly – and this holds true for all action-packed events – always be prepared for the unexpected. When Anatoly Kvochur, one of Russia's leading test pilots, arrived at Boscombe Down in 1991 he kept his audience waiting until the very last minute. As his Su-27P arrived, escorted by two support aircraft, he suddenly selected undercarriage up and roared over 100,000 spectators for an impromptu display of aerial showmanship.

Flying teams are staffed by enthusiasts. Don't be afraid to approach them and ask about the aircraft. If they are not busy they may be only too eager to talk and enliven your video with some anecdotes of air shows gone by. This will make a good sequence to finish on, with some fascinating insights into the world of fliers and flying. □

NOSE DIVES
Here are a few of the most common pitfalls for videomakers with their eyes on the skies . . .

• **PLANE SPEAKERS** – BEWARE OF RAISED OBJECTS LIKE TANNOYS OBSTRUCTING YOUR VIEW OF AIRCRAFT.

• **IS IT A BIRD?** – IF YOU FORGET TO USE THE TELEPHOTO END OF YOUR ZOOM LENS, DISTANT PLANES WILL LOOK LIKE TINY SPECKS IN THE SKY.

• **YOUNG AND IN THE WAY** – REDUCE THE CHANCES OF PEOPLE BLOCKING YOUR VIEW BY CAREFUL POSITIONING.

CAMCORDER CARRIERS

COURTESY OF VANGUARD, OPTEX, AND INTROPHOTO

A STRONG, WELL-DESIGNED BAG OR CASE TO HOLD YOUR VALUABLE CAMCORDER AND ITS ACCESSORIES WILL GIVE VITAL PROTECTION FROM WEAR AND TEAR AND THE UNCERTAINTIES OF THE WEATHER

Your camcorder may have been sold to you without a bag, or was simply in a soft carrying pouch; if so, it is a good idea to invest in a good, sturdy bag or case. As well as offering protection against the weather and all sorts of grime, a good case will substantially reduce the possibility of damage in transit. A camcorder bag must also be able to accommodate any accessories you may need. Equipment like video lights, lenses, filters, tapes, microphones, batteries and even tripods are easily damaged and are better carried in your bag.

There are two basic categories of camcorder carrier: hard and soft. The size and type that you choose will, of course, depend on your requirements. A good dealer will have an extensive range. Whichever model you eventually choose, remember that protecting your camcorder and video equipment is of the utmost importance, so buy the best one that you can afford.

Soft carriers

Soft camcorder bags are usually made from padded nylon and are designed to be slung over the shoulder or carried as a backpack. These bags have the advantage of being extremely light and portable; and the easier your bag is to carry, the more likely you are to take it with you.

Light, soft bags are ideal for family holidays and outings. Although a soft bag cannot guarantee the same amount of protection as a hard case, there are several models on the market that offer greater security than the standard nylon padded bag. These have been treated with a waterproof and fire-resistant finish,

▲ This bag has many external pouches and a pocket for documents in the lid.

OPTEX

▲ The interior of this rigid case is foam-lined and partitioned with movable dividers.

VANGUARD

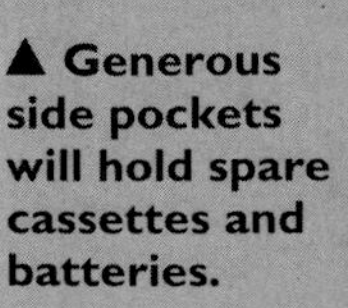

▲ Generous side pockets will hold spare cassettes and batteries.

INTRO PHOTO

CHECK IT OUT

- Meeting your needs: is your chosen bag (or case) big enough to house not only your camcorder but all of your other equipment?
- Portability: are straps wide and padded, to enable you to carry the bag comfortably once it is full of equipment?
- Construction: does your bag have a sturdy base, generous padding and toughened corners to withstand rough handling?
- Adaptability: does the bag have a flexible system of foam partitions or easily removable exterior pockets?

PHOTOGRAPHS BY RAY DUNNS

and are strengthened with moulded plastic corners and bases. If you are worried that your bag may not be waterproof, you can use a shower-resistant spray on it.

However light your bag may seem when empty, it will gain weight rapidly when all your equipment is packed inside. Look for adjustable straps that are wide and padded and that won't dig into your shoulder.

INSIDE INFORMATION

Bag interiors differ greatly, and your choice will depend on how much equipment you have. Choose a bag with a good solid base and look for a variety with thick foam dividers – removable ones are useful as they allow you to adapt the interior to your needs. Mesh pouches in the lid make maximum use of space and are invaluable for those little odds and ends. Many models also have removable exterior pouches and even straps for tripods.

If you are worried about dirt or dust getting in through zip openings, you should consider a top-loading bag. This is fitted with an overlapping dust protection lid – recommended for filming on sandy beaches.

The majority of soft bags are made from nylon in a wide variety of fashionable colours but it is also possible to buy leather bags reinforced with a thick foam lining.

Hard camcorder cases are made of aluminium or a combination of nylon, rubber and PVC cord. Usually they are larger than soft bags, and much heavier but, nonetheless, they are the choice of professionals as they offer maximum protection for delicate equipment. The lids of cases are well padded with honeycomb foam and there are foam sections in the base. As with soft camcorder bags, look for a range of compartments of different sizes and check that the corners and base have been adequately reinforced. Hard cases are usually fitted with a single carrying handle and a detachable shoulder-strap – so they are not too difficult to carry around.

Although quite bulky, and not as portable as a soft bag, a hard case does guarantee the safety of your equipment – and don't forget that it is also strong enough to stand on to give you that added height for any awkward shots.

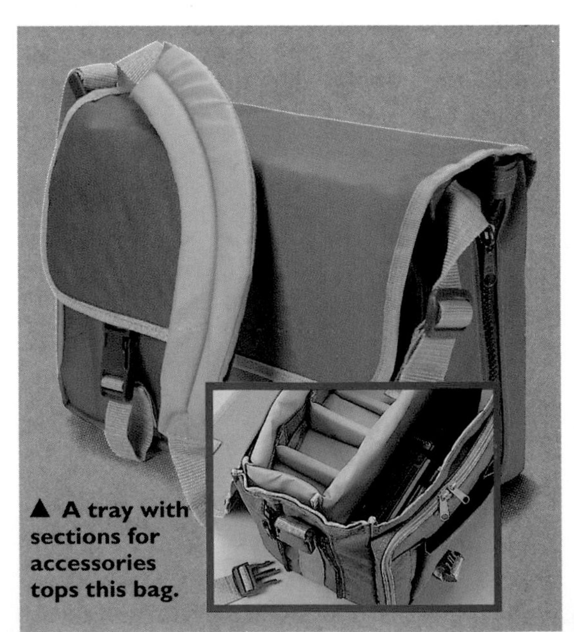

▲ A tray with sections for accessories tops this bag.

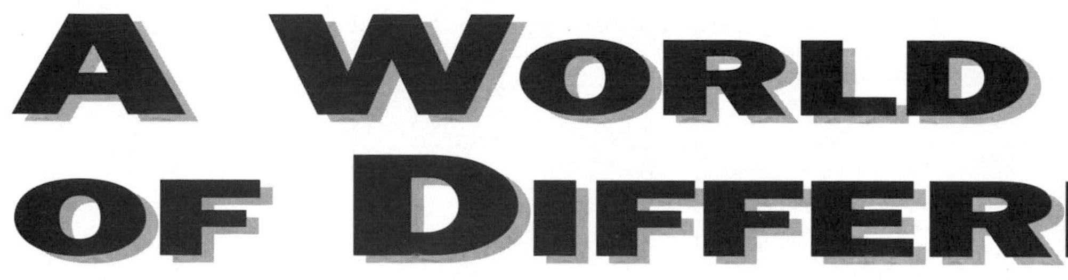

A World of Difference

Copies of family videos make wonderful gifts for friends and relations – but if they live abroad, chances are they won't be able to replay them

The advent of video may have led you to believe that here, at last, was the ultimate, universal, trouble-free and do-it-yourself medium of communication. Reluctant correspondents may have dreamed of kissing letter-writing goodbye for ever and henceforth popping videos in the post instead.

The misery of nagging their offspring to write thank-you notes for Christmas and birthday presents would become just a distant memory. Sending off a tape of the celebrations would be so much more rewarding for all concerned. Family events, such as weddings or christenings, could be shared with those relations and friends unable to attend.

This tempting prospect remains a perfectly practical one – providing the relatives or friends in question live in the same country as you do. Serious problems, however, are likely to arise if they happen to live abroad.

ILLUSTRATIONS BY DAVID ASHFORD

Playback problems

The sad fact is that, though it is easy enough to post your videotapes anywhere in the world, the recipient may not be able to play it back – at least, not in colour. The reason is that different countries operate different colour television systems. The three main ones – identified by the acronyms NTSC, SECAM and PAL – are not compatible with one another. For example if you shoot a video using a camcorder suitable for the UK market, the resulting tape will be 'encoded' in the PAL system. Thus it will not play on a video machine and television set that conform to the NTSC or SECAM systems.

Colour television broadcasting first began in the United States in 1953. The Americans christened their pioneering colour system NTSC, which stands for National Television Standards

◀ For relations who live abroad, videos of their grandchildren and other family members are always appreciated. But you should check whether your TV system is compatible with theirs.

DIAGRAM BY MARK FRANKLIN

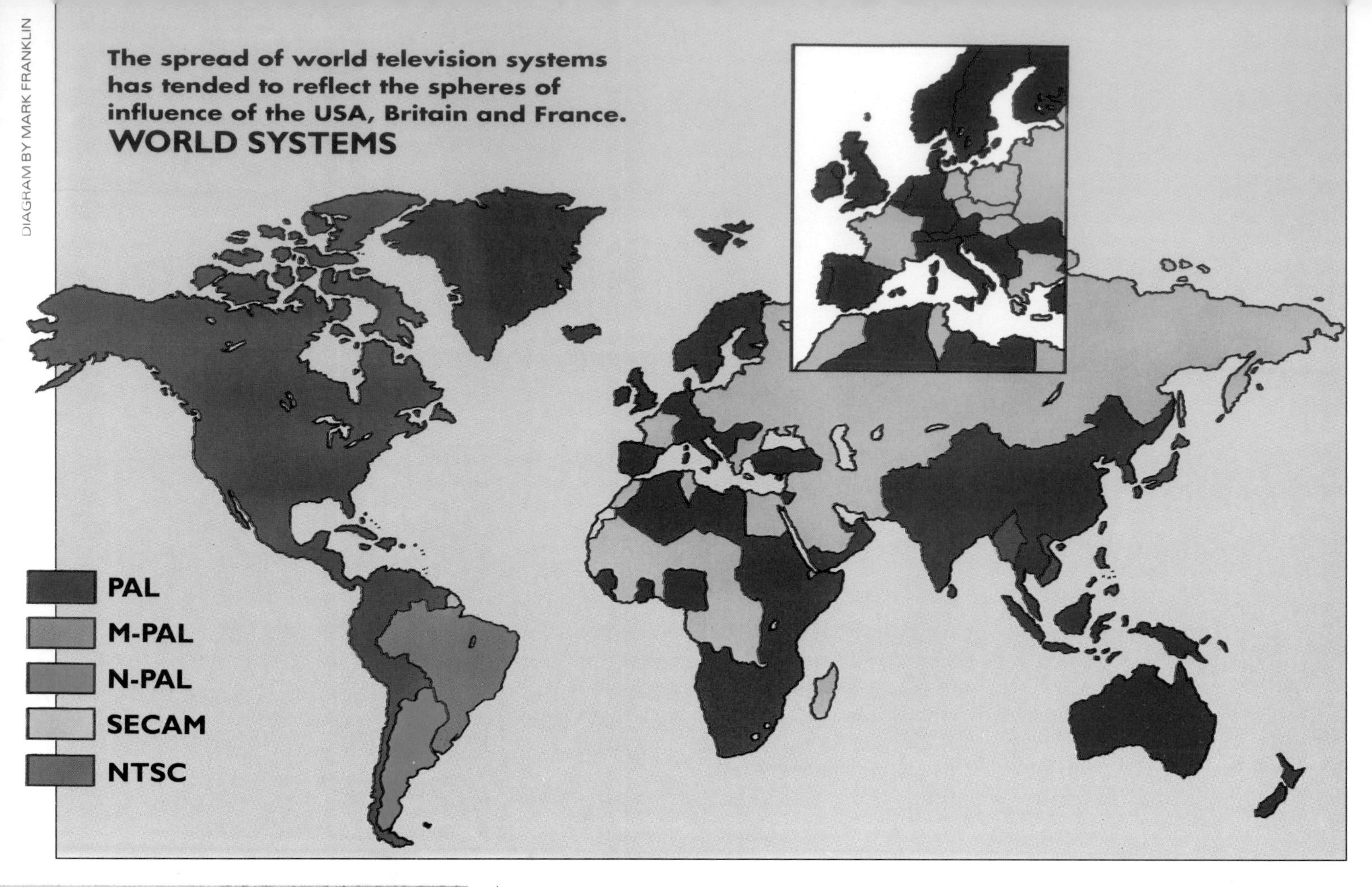

▲ **World systems: there are basically three systems in operation throughout the globe, PAL, NTSC and SECAM**

◀ **Colour broadcasts began in America in 1953, but despite the publicity, they proved to be a disappointment to the public.**

PICTURE: VINTAGE MAGAZINE COMPANY

Committee. However, there were, and still are, problems with the American system. In the early days, the colour varied alarmingly, with faces switching from pink to purple to puce in the space of a single scene. This led some wits to suggest that the letters NTSC stood for Never Twice the Same Colour.

OTHER SYSTEMS

When colour TV was introduced into Britain, more than a decade after the American broadcasts, technology had been improved. PAL (Phase Alternate Line) is developed from the American system but represents a significant improvement (it has been called Peace At Last). This gives a much truer and more stable colour. A purple, puce or green countenance no longer spoils the looks or impugns the sobriety of the performer.

At around the same time that PAL was developed, the French developed their own improved system, SECAM, which also creates a remarkably stable colour system.

The American system has spread through those areas where the United States has had the most commercial and political influence.The British and French systems more or less follow the old colonial lines. Today the cost of standardizing the systems would be astronomical, and given the political jealousies that abound, how could everyone agree on whose system to use?

If you take your own camcorder abroad, the videotapes you produce will be in PAL format and there will be no problem when you want to view them on your own equipment back at home. This does not involve mixing formats. Problems arise only if tapes that you have made in PAL format are to be viewed in a non-PAL country, or you want to view tapes that were made in a non-PAL country. For example, cheap pre-recorded tapes often found on sale in America, may not prove to be such a good buy; when you get them home, you will be unable to play them!

PLAYBACK SOLUTIONS

There are no cheap or easy solutions to these problems for most video users. Even if you want to send videos regularly to relatives abroad, it would be an expensive option to import a camcorder conforming to the standard used in their country, and there could be problems with mains-supply voltage differences and repair service.

No multi-standard camcorders are yet on the market, but it is possible (by special order or from a specialist dealer) to buy a multi-standard VCR, which can record and play back all three types of TV signal in use. It is important to understand, however,

▶ ▼ Another problem is the minor variants of PAL, often called M-Pal or N-Pal. Tapes from these countries are likely to play back in black and white.

NATIONAL TELEVISION STANDARDS

Country	Standard
Abu Dhabi	PAL
Alaska	NTSC
Algeria	PAL
Argentina	N-PAL
Australia	PAL
Austria	PAL
Bahamas	NTSC
Bahrain	PAL
Bangladesh	PAL
Barbados	NTSC
Belgium	PAL
Bermuda	NTSC
Brazil	M-PAL
Bulgaria	SECAM
Burma	NTSC
Canada	NTSC
Canary Islands	PAL
Chile	NTSC
Channel Islands	PAL
China	PAL
CIS	SECAM
Colombia	NTSC
Cuba	NTSC
Cyprus	PAL
Czechoslovakia	SECAM
Denmark	PAL
Dubai	PAL
Ecuador	NTSC
Egypt	SECAM
Finland	PAL
France	SECAM
Germany	PAL/SECAM
Ghana	PAL
Gilbraltar	PAL
Greece	PAL
Greenland	NTSC
Haiti	NTSC
Hawaii	NTSC
Honduras	NTSC
Hong Kong	PAL
Hungary	SECAM
Iceland	PAL
India	PAL
Indonesia	PAL
Iran	SECAM
Iraq	SECAM
Ireland (Eire)	PAL
Israel	PAL
Italy	PAL
Jamaica	NTSC
Japan	NTSC
Kenya	PAL
Korea (South)	NTSC
Kuwait	PAL
Laos	SECAM
Lebanon	SECAM
Liberia	PAL
Liechtenstein	PAL
Luxembourg	PAL/SECAM
Madagascar	SECAM
Madeira	PAL
Malaysia	PAL
Malta	PAL
Mexico	NTSC
Monaco	SECAM
Mongolia	SECAM
Morocco	SECAM
Mozambique	PAL
Netherlands	PAL
New Zealand	PAL
Nicaragua	NTSC
Nigeria	PAL
Norway	PAL
Pakistan	PAL
Panama	NTSC
Paraguay	N-PAL
Peru	NTSC
Philippines	NTSC
Poland	SECAM
Portugal	PAL
Puerto Rico	NTSC
Romania	SECAM
Saudi Arabia	SECAM
Singapore	PAL
South Africa	PAL
Spain	PAL
Sri Lanka	PAL
Sudan	PAL
Sweden	PAL
Switzerland	PAL
Taiwan	NTSC
Tanzania	PAL
Thailand	PAL
Tibet	PAL
Tunisia	SECAM
Turkey	PAL
Uganda	PAL
UAE	PAL
UK	PAL
Uruguay	N-PAL
USA	NTSC
Venezuela	NTSC
Yemen	PAL
Zaire	SECAM
Zambia	PAL
Zimbabwe	PAL

▲ **The Mitsubishi multi-standard VCR – a worthwhile investment if you plan to exchange videos on a regular basis.**

that these machines cannot change the standard: they record and replay whatever type of signal is presented to them. Thus for replay of NTSC tapes the TV set must be NTSC-capable, too. A signal recorded from a PAL camcorder will be laid on tape in PAL format, unrecognizable to an NTSC machine abroad.

The universal

There is an exception: the so-called universal VCR, which contains a complete field-store standards-converter. This machine can be programmed to record or play back in any format, regardless of the standard of the input signal or the pre-recorded tape in use. These special VCRs cost about three times as much as an ordinary one, however, and may not have TV broadcast tuning facilities.

It is now quite common for manufacturers to offer NTSC-replay facility in relatively cheap, home-market VCRs. These models' electronic circuitry converts the off-tape colour signals to PAL standard, but they do not change the picture-scanning rates to match European standards. Many modern TV sets (generally those with screen sizes of 20 inches and upwards) can cope with this. If you consider buying a special-replay VCR, check with your dealer that your existing or proposed TV set is suitable.

Professional conversions

There will always be that occasional special tape – of a wedding, for example – that you will want to send to friends or relatives abroad. In this case the best option is to have the tape converted by a facilities house. Details of these may be found in the small-ads section of video magazines. But be warned – it is an expensive process The tape is converted in 'real time' – one copy of a three-hour tape takes three hours to make and ties up costly machinery. At around £30 an hour, tape conversion involves a large financial outlay.

The future is looking a bit more rosy. There may soon be a worldwide standard for High Definition Television. This will lead to better cinema-style TV, with crisper images, truer colour and better sound for everyone. □

PHOTOGRAPH BY K. HARDI / BUBBLES

Colour control

DOES GRASS LOOK BLUE ON YOUR VIDEOS? DO INTERIORS LOOK YELLOW? IF SO, YOU ARE PROBABLY NOT MAKING PROPER USE OF THE WHITE BALANCE FACILITY ON YOUR CAMCORDER

Anyone who has ever had to struggle with home decorating schemes is likely to have fairly fixed notions about what colour can do to the atmosphere of a room. Blue is cold, red is hot, yellow is sunny and pink is warm. In terms of interior design, pale blue in a chilly, north facing bedroom will make it look about as welcoming as the inside of a deep freeze.

However, as far as the camcorder is concerned, blue is hot stuff and red is cool. The reason for this is that the human eye and the camcorder do not work in the same way.

The tricks that light plays on colours do not matter too much as long as it is your eye that is doing the seeing. The human eye is backed up by the brain which is programmed to adjust to changes in light and colour

LEARN HOW TO

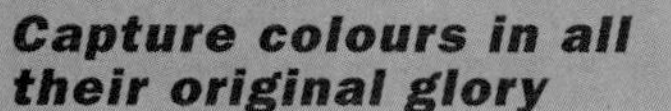

- ***Capture colours in all their original glory***
- ***Cope with varying light conditions***
- ***Create atmospheric colour effects***

COLOUR TEMPERATURE SCALE

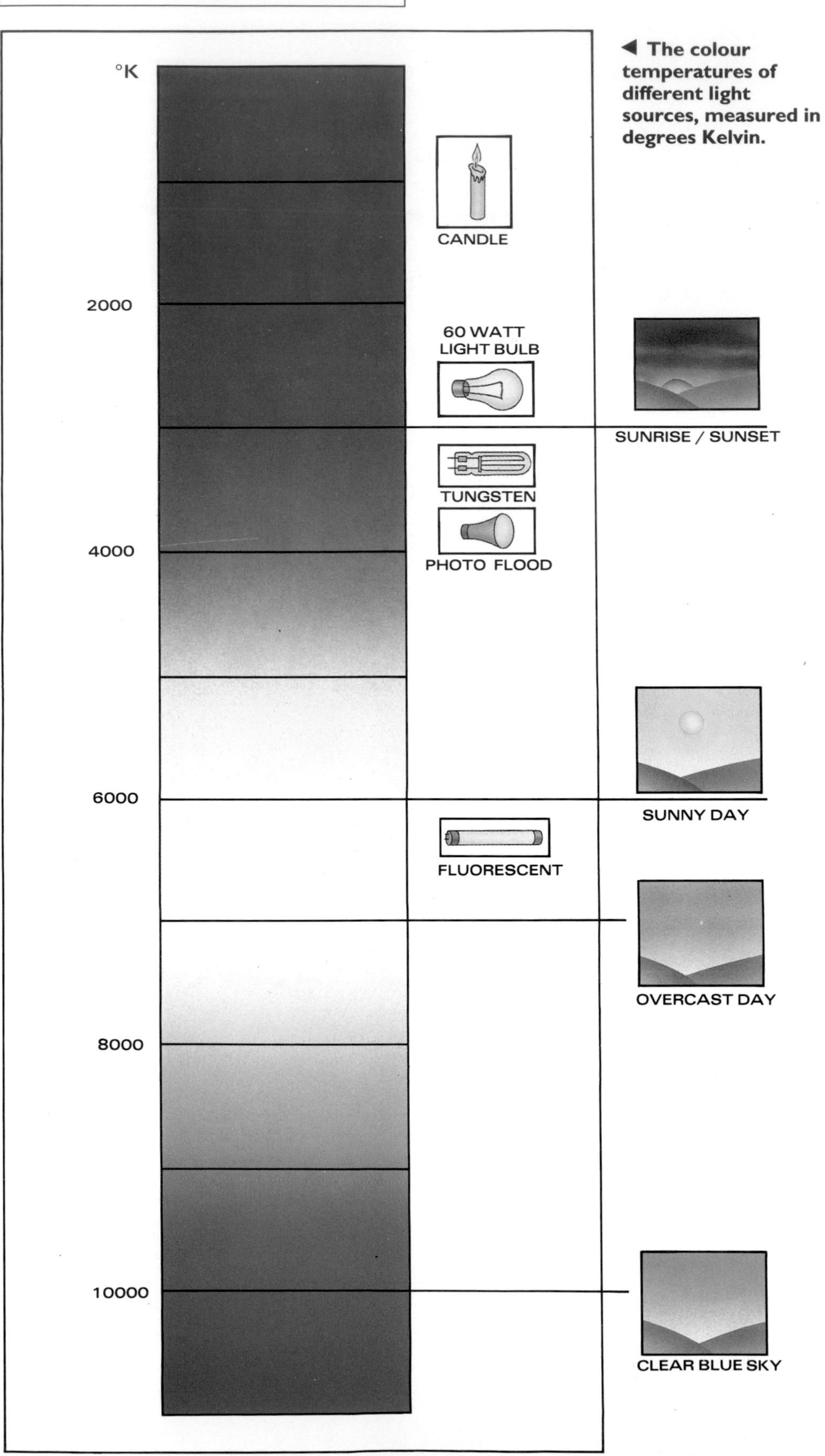

◀ **The colour temperatures of different light sources, measured in degrees Kelvin.**

DIAGRAM BY MARK FRANKLIN

where necessary: a green apple will still look more or less the same green whether it is seen in daylight or indoors by candlelight.

However, all sorts of peculiar things will appear to happen to that green apple and its surroundings if it is videotaped under a variety of different light conditions. For example, bright daylight will introduce a pale blue tinge, while sunrise or sunset may lend it a slight rosy or orange glow; tungsten (indoor and artificial) lighting will bring a distinctly yellow hue to the fruit and a fluorescent tube may well give the apple and its surroundings a slightly bluish cast.

Colour temperature

This rather confusing discrepancy between what the eye sees and what the camcorder sees relates to the nature of light and its relationship to what is known as colour temperature. An understanding of colour temperature is important because it will affect the colour quality of your video. If you do not take account of it, the finished product is likely to be far from pleasing to the eye.

Isaac Newton demonstrated that pure white light is not white at all, but a complicated mixture of lots of colours. This range of colours runs from violet through blues, greens, yellows and oranges to red and is called a spectrum. Colour temperature is a way of expressing the colour bias of any given light source, and is measured in units called Kelvins.

CAMERA SENSE

MIXED LIGHTING

A difficult shooting situation to deal with is one that involves mixed light. This is often encountered when you are shooting indoors with daylight from a window supplemented by a lamp. Automatic White Balance tends to give muddy colours in this situation. You can solve the problem by the use of a blue photographic gel on lamp bulbs to raise their colour temperature to that of daylight. An alternative is to use a special light bulb with a daylight colour balance.

PHOTOGRAPHS BY LYNDON PARKER

▲ The blue cast given to an outdoor scene by an indoor setting.

▶ The correct colouring given by an outdoor setting.

▲▶ A daylight setting produces a yellowish cast over an indoor scene, whereas the appropriate indoor setting ensures correct colour rendering.

Throughout the day there are numerous changes in colour temperature. Seen via the camcorder's lens, light at sunrise will have a warm pink or yellow colour bias and register at around 2000 to 3000 degrees Kelvin. Colour temperature peaks around noon at between 6000 and 7000 degrees and will be white or pale blue in colour. Summer shade will have an even bluer tinge, at around 8000 degrees. As the day wears on, the colour moves through pink to the oranges of sunset as the colour temperature cools again.

Artificial light is also measured in colour temperature terms, but unlike daylight does not contain a full spectrum of colours. This can have a dramatic affect on the way things lit by it appear to the camcorder.

Candlelight is around 1000 degrees and gives scenes a reddish cast. Domestic light bulbs have a colour temperature of 2800. This is close to the 3200 degrees of all quartz battery and mains video lights – the tungsten/indoor setting manufacturers fix for camcorders. Photoflood bulbs are slightly higher at 3400 degrees. The colour temperature of all three sources is close enough for differences to be only minimal. Interior scenes lit solely by domestic bulbs can appear slightly warmer when recorded using the tungsten setting. However a fluorescent tube is much warmer, at between 6000 and 7000 degrees , and gives a white or pale blue light.

WHITE BALANCE

Because of these differences in colour temperature, every current camcorder has an Automatic White Balance feature designed to

CAMERA SENSE

ENHANCERS

If you find that your footage has an unpleasant colour cast, all is not lost: a machine called a video enhancer will help you to correct colour imbalance. Most enhancers have three types of control: colour saturation, brightness and/or contrast, and colour correction. The latter will enable you to put a colour cast over any frame in your video to correct imbalance.

compensate for changes in ambient lighting conditions, such as when you are shooting indoors and then move outside for exterior scenes. Without it there would be no continuity of colour, and the green apple could swiftly change from a natural green to a blue or jaundiced yellow, depending on its surroundings and the type of light source.

MANUAL AND AUTOMATIC

Most camcorders' Automatic White Balance feature a Manual override. Slightly older models, however, may possess Manual adjustment only. These latter models can be a little frustrating: you need to adjust the White Balance every time the light seems to change, and in the heat of the moment, it can be difficult to remember to do this. On the other hand, the sun passing behind a cloud at an outdoor shoot, or movement between any indoor and outdoor locations, should pose no serious problems to a camcorder with Automatic White Balance.

Camcorder booklets will instruct you how to use the Manual or Automatic White Balance on your model, but may not explain the differences

IMPROVE YOUR SKILLS

In the Mood

Although correct lighting is important for the average shot, it may not be suitable if you want to endow the scenes you are shooting with a particular mood. To get the right effect for these, practise using a Manual White Balance setting that seems entirely 'wrong' for the circumstances.

◄ A cold dinner . . . Here candlelight and artificial ceiling light are faithfully reproduced using the correct indoor setting.

▼ A daylight setting used indoors has now cast a warm tinge over the scene, giving it an intimate glow.

◀ The daylight setting retains what little warmth there is in the scene.

▶ An indoor setting brings out the blues in the outdoor scene, making it look as cold and grim as it probably felt at the time!

▶ This church tower looks charmingly picturesque on a sunlit day as rendered by the daylight setting.

◀ Using an indoor setting's bluish cast, however, you can give the tower an air of gloom and melancholy.

between the two. This knowledge is useful if you have an Automatic White Balance with Manual override.

The Manual White Balance feature has two settings – for daylight and tungsten (indoor) light. It is important to remember to set your White Balance at each change in location or light source. The usual drill is to fill the viewfinder with a close-up of a white card or sheet of paper. (It is also possible to get a white lens cap that will save you having to cart a white card around wherever you go.) The next step is to defocus and then to press the White Balance switch for five seconds. The White Balance for those lighting conditions is now set. This means in essence that all colours

Q

Which manual colour temperature should I use to video in our office? It has both fluorescent lighting and plenty of daylight.

Use the daylight setting. The colour temperature of fluorescent light varies between 6000 and 7000 degrees. This is close to a camcorder's daylight setting of 5600.

A

will remain true while the light source remains the same. Bear in mind that it is all too easy to forget to reset the White Balance because the image in your viewfinder is in black and white, and there will be no indication through the viewfinder of any lighting or colour problems.

Problems with automatic

Automatic White Balance functions, on the other hand, rely on readings taken by a small white panel built into the body of the machine. However, even an Automatic White Balance feature can present you with some unpleasant surprises, despite the claims of manufacturers. It cannot

SPECIAL EFFECTS

If you wish to make the most of a gloriously romantic sunset, use the daylight setting on your camcorder. This will bring out the deep reds and oranges of this delightful spectacle. If you are shooting an ancient monument or the picturesque ruins of an abbey or castle in daylight and want to achieve an ethereal quality, use the tungsten setting. This will lend an air of remoteness and mystery to the scene by producing a silvery-blue cast over the frame.

PHOTOGRAPHS JANET & COLIN BORD / FORTEAN PICTURE LIBRARY AND M. DECET

cope with a mixed light source, such as daylight and tungsten lighting a scene, for instance. It will opt for something in the middle and make a real mess of things. When faced by mixed lighting, a camcorder with a White Balance lock (a Manual override) will help you to achieve the right effect. By framing the main subject matter and locking on to it with the White Balance feature, you can at least ensure that this area will be coloured correctly.

In short, total reliance on the Automatic White Balance can be quite foolhardy. It is essential to test new equipment thoroughly and to experiment with it before you trust yourself and it with an important, one-off shoot, such as a family wedding.

Monitoring colours

If possible, keep a check on colour quality with a monitor or a portable colour television. Take note of the key colours, such as flesh tones or grass, which must be correctly represented.

The wrong White Balance setting will not always result in disaster. In fact, an 'incorrect' colour temperature can be put to creative effect. Using the daylight setting indoors, for instance, produces a warm, cosy glow – ideal for occasions such as a family Christmas. Similarly, using an indoor setting outdoors creates a cool, crisp atmosphere. On a slightly overcast day, when there are few shadows or strong light sources, this gives the effect of moonlight.

Tint experiments

These effects are easy to produce. Simply set the camcorder's White Balance in one location, and move to another to shoot.

Some camcorders register light conditions through a window on the casing. With the Automatic White Balance switched on, the sensor can be covered by a transparent coloured paper and the picture will take on a complementary tint. With other camcorders, a coloured card can be placed in front of the lens when setting the Automatic White Balance to produce a similar effect. Better still, some camcorders have a control that enables you to choose a warmer or colder colour cast at will. □

COLOUR BLIND

When your viewfinder gives only a black and white image, it is easy to set the White Balance incorrectly . . .

• **Got the blues** – Using the indoor setting in hazy sunlight can give whites a bluish tinge

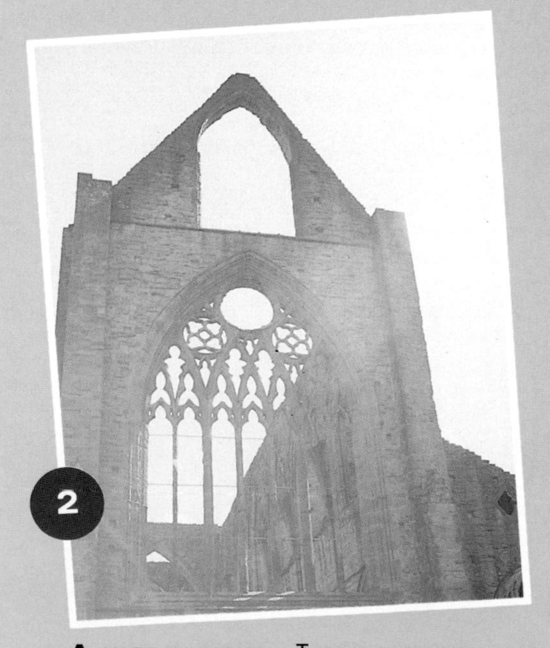

• **A shot in ruins** – The washed-out effect resulted from failure to recognize the depth of blue on the abbey: the indoor setting could have been used to bring it out

Dubbing music

ILLUSTRATIONS BY ANDREW HARRIS

IT'S NOT ONLY PICTURES THAT TELL A STORY – A DRAMATIC MUSICAL SOUND-TRACK WILL ADD INTEREST AND IMPACT AND CAN BE EASILY DUBBED ON TO YOUR VIDEOTAPE

ou've shot your video and edited the images together and now all your film needs is a sound-track. You will have recorded some sounds together with your pictures but your microphone may have picked up some unwanted noise and, when editing and cutting shots together, you will probably get some jumps in the sound-track. This isn't to say that you will not be able to use the original sound but you will probably want to pick out only those effects that enhance your storyline.

You can add to the sound-track by dubbing in other material and music is an ideal way of both unifying the visual images and creating the right atmosphere for your video. A video taken on holiday might be a fascinating record of a city's architecture, but it will probably contain varying levels of traffic noise that will only be distracting. By adding, for instance, a piece by Handel to your film of a day out in Bath, you can reinforce the images of a serene and classical city.

MUSICAL ATMOSPHERE

Music can set the tone for the whole of a film and you have only to listen to the sound-tracks of major movies to understand the impact that the right music can make. As well as underlying the mood of a film, music can increase atmosphere and intensify drama. Moreover, it bridges cuts, links sequences, adds pace and helps the viewer understand what is going on. Action sequences need dramatic, fast-paced music, while more reflective moments will call for gentler tunes.

When you make your choice of music, try not to use well-known themes as they can easily evoke strong associations that may

▲ It does not necessarily take a vast array of highly technical equipment to dub music on to your video. Only simple skills are needed to produce a quality sound-track.

▲ **Classical music can be ideal to record on to your video sound-track and there is something for any mood or occasion. Ballet music may be an obvious choice for a shot of dancers but it might also be well suited to more unusual situations.**

overpower the visual storyline. It is also worth remembering that musical recordings are subject to copyright restrictions. Copyright is limited to 50 years after the death of the composer but even classical music may be effected by the law since the

SOUND ADVICE

PREPARING THE SOUND-TRACK

If you are planning an elaborate video, jot down your music ideas before shooting – good sound-tracks need careful preparation. But if your video is simply a record of your holiday, you need not think about the music until you have seen the tape. Sometimes the length of the music will dictate the scene's length, so the picture will need to be edited to match the sound. Check the time read-outs on your camcorder and VCR to determine the length of scenes.

arrangement and performance of a piece will by copyrighted. Although you need not worry about breaking the law if your video is strictly for personal use, if you plan a public showing, clear any use of material protected by copyright with the appropriate authorities. A solution might be to use specially recorded, copyright-free music, and there are mood music publishers who supply music on tapes, albums and compact discs. You can find advertisements for some of these suppliers in specialist video magazines.

COMPOSING YOUR OWN MUSIC

Alternatively, you could try composing your own score. A keyboard and a sequencer can enable even the most amateur of musicians to put sounds together, store music note by note, and insert or delete material at a different stage. Add a tape recorder and you are well on the way to having a home studio.

Little equipment is needed to achieve good musical dubbing and careful planning can help produce a highly effective sound-track. The simplest way to put new sound on to your videotape is with the Audio Dub facility of a camcorder or video recorder. This allows new sound to be recorded without affecting the picture. Audio Dub is handy because it can be used to supplement some, or all, of the live sound in parts of the video when this is not adequate on its own.

Before you begin, take stock of the specification of your camcorder and video recorder. If they are VHS mono, then the new sound will simply dub over the already recorded sound, wiping it out. However, 8mm sound and VHS stereo are fundamentally different.

VHS STEREO FORMATS

VHS stereo formats have three sound-tracks – one mono and two stereo. Unlike the mono track, the stereo sound-tracks cannot be altered without affecting the picture, because the sound and video signals are inseparable. But this has no bearing on the mono track, on which you can record your choice of music. Nevertheless, a word of caution is needed. The stereo and mono tracks cannot be balanced without re-recording and mixing them, so control the music level from the first so that it doesn't dominate the stereo tracks. Do a test to determine the right sound level. Of

CAMERA SENSE

WORKING FROM A COPY

If you make a mistake while adding music to your video, you might accidentally record over the original picture or sound, or both. The solution is to work with a copy. Going to another generation doesn't lose much quality and keeps your original video safe. Should you make a mistake with a scene, you can always re record that scene from the original.

course, if you have not recorded any sound, balance will not be a concern.

Camcorders and video recorders in 8mm format all have FM audio systems. The sound and video signals are inseparable and cannot be altered without affecting the picture. The exceptions are machines with PCM (Pulse Code Modulation), which additionally record stereo tracks separately from the video signal – but only a few of these have been produced and they are expensive. Sound-tracks on standard Video 8 or Hi-8 machines can be mixed or replaced, but only while copying or editing the video signal on to another tape. It might be worth your while to buy an inexpensive Audio Mixer which you can use to blend sound from the original tape with more than one other source. Mixers generally have two or three controllable inputs which enable you to add music to your sound-track while adjusting the level of what you are recording.

PLANNING AND TIMING

After getting to grips with your equipment, turn your attention to the sound-track itself. Choose which music you want to go with each scene and prepare a simple dubbing chart. This is a visual plan of where music starts and finishes in your video. Concentrate on accurately timing your music so it matches the visuals and make sure you know where in your video you want each selection to go.

The simplest, but crudest, way to dub music is to plug a microphone into the external mic socket of the camcorder or video recorder. If either has manual Audio Level Control, you can adjust the volume to suit the subject. Then position the

▼ Why not pick something like Tchaikovsky's *Swan Lake* to accompany your film of a rugby match? You can give your video a humourous note if you take music with strong associations, such as ballet music, and combine it with highly contrasting images.

microphone near one of your stereo system's speakers. Next, rewind your videotape to the point where you want the dub to begin. Find the precise point in your music source – a tape, for example – where your selection starts, and wind back an extra second or so and press Pause. Carry out the same process for the source videotape and engage the Audio Dub and the Pause modes. Release the Pause on the video unit and then on the audio source. When the music has finished, press Pause on the video unit and then stop the audio. There is a certain amount of trial and error involved here, but with a little bit of time, practice and patience you will soon be able to line up both audio and video to your satisfaction.

Using this basic method can work well, except that the microphone can pick up unwanted sounds and the Automatic Gain Control on the camcorder will tend to level out the volume of the music. However, for a more polished effect, connect a lead direct from the audio source to the Audio-In socket of the camcorder or video recorder. This source can be a tape recorder, a compact disc or record player – or even another video recorder. Using a direct connection in this way, rather than a microphone, you will not accidentally pick up unwanted additional sound.

If you get used to these simple techniques you will soon be able to put together a sound-track for your video that is both interesting and enlivening. As you get more and more proficient with dubbing music you will be able to move on to more sophisticated audio dubbing, such as combining music and commentary. □

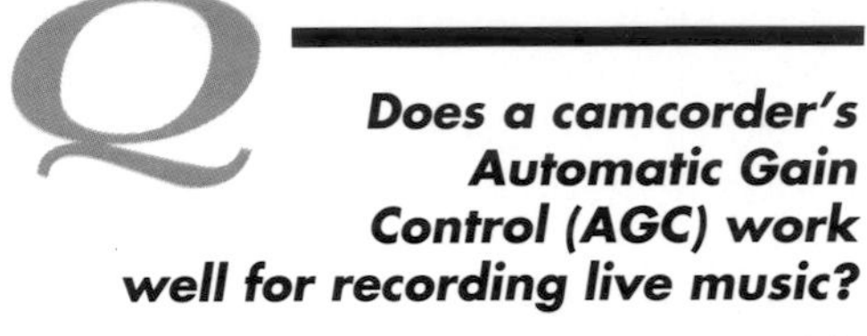

Does a camcorder's Automatic Gain Control (AGC) work well for recording live music?

Using the AGC for live music is a gamble because it works best with constant sounds. The AGC averages out changes in sound level and during a lull in a song it will increase the background noise to keep the same level as the music, picking up sounds like talking or coughing.

A

▼ **When you are editing your finished film an accurate record of the length of each scene will help you match your soundtrack to the images. A simple stopwatch can help you time both the action and the music for you video.**

Beside the seaside

A VISIT TO THE SEASIDE COMES HIGH ON THE LIST OF MOST CHILDREN'S FAVOURITE TREATS. PADDLING IN THE WAVES, EATING ICE-CREAMS AND GAMES ON THE SAND ARE JUST SOME OF THE LIKELY INGREDIENTS OF A DELIGHTFUL FAMILY VIDEO

The essence of a day on the beach with the family is its relaxed and unstructured nature but, to create a memorable video, you will need to plan well ahead. Unlike recording events indoors, you will, of course, have little control over the lighting, background noise and weather conditions by the sea. In addition your family will not take kindly to being bossed around purely for the sake of getting good shots on tape, so you will have to content yourself with shooting events as they happen. Even more important, taking a camcorder to the beach without proper safeguards breaks all the basic rules of equipment care by exposing it to the ravages of sand, salt spray and, quite possibly, your children's ice-cream-covered fingers.

The lie of the land

The degree of detailed planning you can make in advance depends on how well you know the location. The advantage of shooting while you are

PHOTOGRAPHS BY ANTHONY ROSE

staying in a resort as part of your annual holiday is that you soon get to know the best beaches, the times of high and low tide, ideal camera positions and any special scenic features you might want to include, such as sand dunes, cliffs or rock pools. You are also far more likely to be aware of the timing of the tide – crucial in areas where the water-line is very high.

However, it is still quite possible to make effective plans for a day at an unfamiliar resort. Good maps differentiate between sand and shingle beaches, and brochures will show you the general layout of the place. Try to explore promising locations well in advance A telephone call to a local tourist information centre will give you tide times and details of any special events you can incorporate into your video, too.

PLANNING CHECKLIST

- **Safeguard your equipment from sun, sand and spray. Enclose your camcorder in a special waterproof housing**
- **Bring a tripod, and filters for shooting in bright light**
- **Remember to take extra tapes and a fully-charged, spare battery pack**
- **Check tide times and the possibility of special beach events before you go**
- **Choose your 'base' on the beach to avoid long treks to natural features such as sand dunes and rock pools**
- **Check the fastenings of your equipment bag close fully**

▲ Before the family arrives, record a leisurely panning shot of the beach to set the scene. Later you may wish to overdub a burst of music at this point to emphasize the unspoilt grandeur of the setting.

PROTECTIVE MEASURES

Before you set off, make sure you are confident of operating all your camcorder's controls manually. While automatic exposure is an advantage when you're concentrating on the action, strong sunlight can cause the iris in the lens to close down and lead to under-exposure of a backlit subject. It is a good idea to fit your lens with a UV (Ultra-violet) filter, to protect it against the elements and to guard against glare from sea and sky. In addition, an ND (Neutral Density) filter, which reduces the amount of light falling on the lens will be useful on a sunny day to prevent over-exposure.

▶ The peaceful atmosphere is about to change as Mum and the kids arrive on the beach. Try for a fairly low-angle shot as they come towards you over the sand.

SEASIDE SOUND-TRACK

Much of the joy of a visit to the seaside is the wonderful symphony of noises all around you. Gulls calling as they wheel in the sky, shingle shifting to and fro at the water's edge and even the excited barking of a dog frolicking in the waves can all add atmosphere to your soundtrack. But you will need to prevent these evocative background sounds from becoming an intrusion by making sure you get a visual record of them, too.

▲ 'Gently does it' – a sequence of Dad trying to change into his swimming trunks without a slipping towel revealing all is a sure crowd-pleaser. Get in close to record his brave attempts to laugh the whole thing off!

You can create an attractive opening title for your video simply by writing in the sand with a stick or by arranging shells or pebbles into letter shapes. This idea can also be employed as a closing shot, showing your title being gently erased by the incoming tide.

If you have decided to shoot your day at the beach in the middle of your annual holiday, you will already be familiar with the likely pattern of the day and can use this as the structure for your video. If visiting the seaside just for the day, you will have to take your chances as they present themselves but at least you will have the added bonus of extra enthusiasm from the children.

This excitement can make a tremendous opening shot, especially if accompanied on the sound-track by gasps and shouts of 'The sea! The sea!'. If you can zoom out and then pan slowly in the direction of the children's gaze for a wide establishing shot of the beach, so much the better.

QUICK-CHANGE ARTISTE

Few sights on a beach are more comical than the acrobatics necessary to change from ordinary clothes into swimming costumes. The embarrassment of a member of the family being

▲ **Time for a paddle – after an establishing shot showing Mum taking the children into the waves, Pause and change position. Shoot from a low angle to capture expressions.**

recorded clutching a towel while balancing precariously on one leg adds to audience enjoyment. But select your subject with care: younger children won't use a towel in the first place and older children may well resent the loss of their dignity. It may be better to volunteer yourself as 'victim' and hand the camcorder to another member of the family.

In the swim

After this, take your lead from the children. As soon as they have changed into bathing costumes, most kids will run straight for the waves; if you can get there first, you should be able to video them as they sprint to the sea and capture their startled expressions as they adjust to the temperature of the water.

First strokes

Children splashing about in the sea makes a great sequence – especially if they have an airbed, a beach-ball or swimming rings to play with. Don't forget to enclose your camcorder securely in a waterproof case. For safety's sake, you will also need another adult either in shot or watching nearby just in case of accidents. This is especially important if one of your children is taking the plunge for the very first time – as much of a joy for any parent to record on video as their first steps on dry land.

▲ **Always try to get as close as you can to the action in order to involve the viewer in the scene – sequences composed only of long or wide shots distance the viewer and can be a little boring to watch.**

If the water is clear enough, or there are rock pools nearby, make sure you get a sequence of children looking for fish and crabs. Life underwater holds great fascination for young ones. By crouching low, you may even be able to get a shrimp's-eye-view of the children.

After all this waterplay, it will be time to dry off. This gives a good opportunity for smiling close-ups of

ENEMY TERRITORY

Even on still days, a beach is a hostile environment for a camcorder. You will need to buy a sand and waterproof housing and practise using it before arriving at the beach. Alternatively, you could buy or hire a special sports camcorder.

BOBBING ALONG

To get the most out of your waterproof camcorder case, try for some semi-underwater sequences of your subjects. Lay in the sea and shoot them coming towards you through the water, or crouch in a rock pool to catch a fish's-eye-view of them. Allowing the lens to bob up and down with the rocking of the waves will enhance the effect but let the lens clear the water enough for your subjects to come into focus before dipping it partly under the surface again. The water is unlikely to be clear enough for you to shoot with the whole lens submerged.

PHOTOGRAPHS BY MIKE HUNT

▲▶ After splashing about in the sea for a while most children will be in the mood for a rub down and a quick cuddle to warm up – another ideal opportunity for a telling close-up.

the children wrapped in towels and in the arms of a parent. Go right in for a big close-up of child and parent; this will help change the visual pace of your recording by contrasting with the wide open spaces of the beach.

Warm the children up again with some beach games. Football, cricket, rounders, anything that lets them run around for the camcorder. You'll need to stand back and use a wide angle shot to catch most of the action in context. But especially for a game such as cricket, you can also zoom in for telling close-ups of the players preparing for action.

OTHER POINTS OF VIEW

Don't forget while all this is going on to pass the camcorder around the family. Videos in which Mum or Dad do not appear at all give a very one-sided look to the day. If your children are old enough to support the camcorder's weight, let them have a turn, too. Slightly wobbly shots of your efforts to score a goal or hit a six can all add to the comic atmosphere. Another way that you can ensure the entire family appears in your video is

to set up the camcorder on a tripod, frame up a wide shot, press Record and video everyone sitting together having a picnic lunch.

Construction work

No video of a day at the beach would be complete without a sequence showing the children making a sand-castle or two. And if you pace your recording, you can get a great sequence. Sand going in the bucket, scraping off the excess and tipping the bucket over makes a nice start. From there, just record every second or third 'tip out' and then show the children decorating their creations. This should provide an opportunity to move in for unguarded close-ups of the children as they carefully put shells or pebbles in place.

Digging around in the sand does wonders for children's fantasy play – they are quite likely to become so engrossed in what they are doing that they will be completely unaware of the video camera. If you are accomplished at shooting blind, hold your camcorder down low at your side to allow the viewer to see things from their point of view – it is a common error when videoing children to shoot from too high an angle. Move in as close as you can without distracting them and pick up as much of their conversation as you can.

▲▶ Kids become so engrossed making sandcastles that, for once, they'll forget you are taping their every move.

Camera Sense

SHEDDING LIGHT

The quality and strength of light available outdoors can vary greatly – and what is good for beachgoers may not be the best for your recording. Strong sunlight can lead to overexposed fleshtones and to faces squinting against the glare. You may get best results when there is some cloud cover. Shooting early in the morning or late in the afternoon can also produce good lighting conditions.

▲ **Burying Dad in the sand is great fun for the kids and should make an hilarious sequence.**

▶ **A game of cricket will help to warm everyone up after a dip in the sea. Take a shot of Dad throwing the ball, then Pause and change your position as the 'batsman' hits out.**

While the children are absorbed, don't forget to record Mum or Dad relaxing. Sneak shots of a member of the family snoozing with their mouth wide open and apparently without a care in the world, always have an amusing quality and somehow sum up what a holiday or trip to the beach is all about – getting away from it all. If you are planning to edit your tapes at home, this sort of shot may prove very useful as a cutaway to insert, for comic effect, into an action-packed sequence, such as a beach game.

Demolition derby

Create a dramatic contrast to the slow tempo of building the sandcastle by challenging the kids to knock the lot down as fast as they can. To avoid flying sand, stand back and shoot on wide angle, zooming in on their trampling feet for added effect.

If you haven't already faced your children's demands that they be bought ice-creams, they are sure to come towards the end of the day – and what a super sequence this can make, especially where younger children are concerned. Move from a wide shot to a close-up as their faces grow messier and messier – and remember to keep your camcorder well away from sticky fingers.

While the children demolish the remainder of their ice-creams, take some time to shoot some of the other things going on around you. Sailboarders, surfers, sunbathers, sailing dinghies, light aircraft and ships on the horizon are just some of the cutaway shots you can edit into your video later to change the pace or alter its balance and mood. It is all too easy to become so preoccupied with your own family's activities that you completely forget to capture on tape some of the unique atmosphere of the seaside resort itself.

Afterwards, it will probably be time to go home. A brief sequence of shots showing the family packing up

▲ **Get right down to the children's level for a close-up sequence as they tuck into their ice-creams – for once, the messier they get, the better it will look.**

▼ **If shots of packing up are not to your taste, fading out on a close-up of a happy face is one way of ending your video on a carefree note.**

all the beach things and carrying them away will make a suitable closing scene to your seaside video. Move in close to record hands picking up bags, buckets and spades, then tilt slowly up and zoom out as your family treks away from the sea. Finally, you could fade out on a shot of the sea lapping over a sandcastle or an end title written in the sand.

SUNSET SPECTACLE

If you are still on the beach at the very end of the day, try for a shot of the sun setting. Many people regard this kind of image as something of a travel-brochure cliché – but if it's a spectacular shot, there is absolutely no reason not to employ it to end your tape. Remember to set your camcorder's White Balance to daylight: this will bring out the reds and oranges in the sunset.

Closing shots such as these always have a melancholy quality however good the day has been and will nicely complement the excitement of your opening sequences. □

SEA OF HEARTBREAK

Sea, sun, sand and children can spell video vexation for the unwary camcorder user . . .

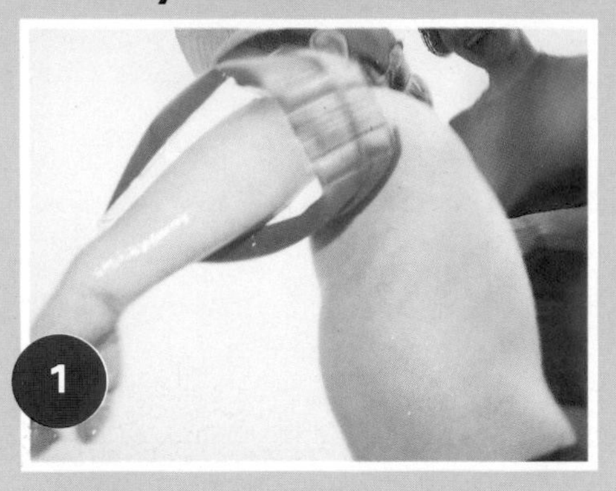

• **OVER-EXPOSURE** – THE SCENE HAS PROVED TOO BRIGHT FOR THE AUTO-IRIS. USE A NEUTRAL DENSITY FILTER

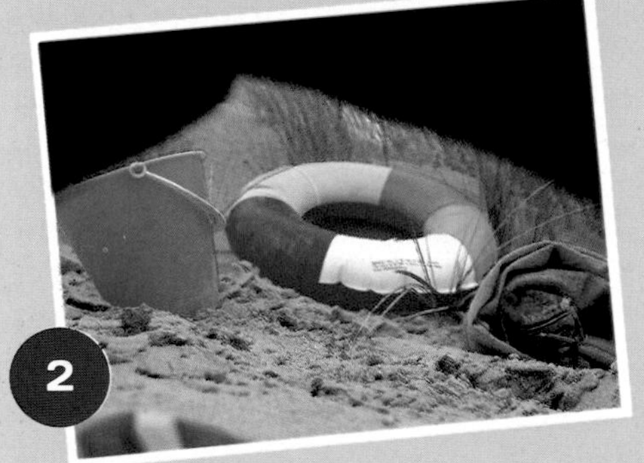

• **BAG OF NERVES** – IT'S SURPRISINGLY EASY TO TURN ON A CAMCORDER BY ACCIDENT WHEN YOU THINK IT HAS BEEN SAFELY STOWED AWAY IN THE BAG

• **STICKY FINGERS** – BE SURE TO KEEP YOUR CAMCORDER'S LENS WELL AWAY FROM SUNTAN LOTION, WATER AND SAND

Battery Pack Lamps

Shed a little light upon your subjects with a clip-on, battery-pack video lamp. Small, lightweight and portable, you'll be surprised at the difference it makes to the quality of your videos

Unlike a still camera, camcorders can operate in extremely low light conditions. But there comes a point where the picture quality deteriorates to such an extent that some additional light source is necessary. It is not always possible to increase the ambient light of the scene (at a bonfire party or a wedding reception for example) so the solution is to provide the light yourself with a clip-on, battery-powered video lamp.

All video lamps consist of a small, efficient quartz bulb and reflector, contained in a small plastic case. They are fixed to the camcorder in several ways: some slide on to the

▼ Battery-pack lamps can improve the quality of your videos by lightening murky indoor scenes and improving dull outdoor sequences.

◀ **An accessory-shoe-mounted lamp with its own chargeable battery. The two bulbs can be switched on individually or together to provide 10, 20 or 30 watts of power. The base is hinged to angle the beam and avoid shining the light in your subjects' eyes.**

camcorder's shoe, or on to the battery housing at the back. If your camcorder does not have a shoe, you can buy a mounting bracket which screws in to the tripod housing on the bottom of the machine.

Lighting options

All video lights are designed to provide you with extra light over a limited area. Just how bright that light is depends on the wattage of the lamps. A 10-watt bulb is perfectly sufficient for fill-in lighting – where you want to illuminate a person's face against a dull background, for example, but a 20-watt is better as an all round 'starter' light. It does not necessarily follow that a 20-watt bulb is twice as bright as a 10-watt – its brilliance depends on the efficiency of the bulb and the reflector system. Some battery pack lamps comprise two bulbs – a 10-watt and 20-watt – in individual housings. These can be used independently, or combined to give a full 30 watts of light. For lighting above 30 watts, you will need to invest in a video light. These powerful, scaled-up versions of battery pack lamps provide between 100 and 1000 watts of power. These large, hand-held or

BUBBLES/PAUL HOWARD

▲ **Shooting outdoors at night becomes easy with a battery lamp – but don't be tempted to use it in the drizzle or rain for safety reasons.**

Camera Sense

ACCESSORY SHOE OR MOUNTING BRACKET?

Some camcorders are built with an accessory mounting on the top to which you can fix lamps, microphones or other supplementary equipment. This is known as an accessory shoe. If your camcorder does not have one, you can still attach any extra accessories you need by means of a separate video mounting bracket. The bracket screws into the tripod fitting at the bottom of the camcorder, and has one (and sometimes two) equipment mounts at the opposite end. Their special angled design, with an integral grip, offers extra stability when filming. There are various bracket models available – make sure you choose one that is fitted with a screw *and* a housing at the base, so that you can still fit a tripod to the bottom of the camcorder when the bracket is in place.

LIGHTS COURTESY OF VANGUARD AND OPTEX

SPECTRUM COLOLF LIBRARY

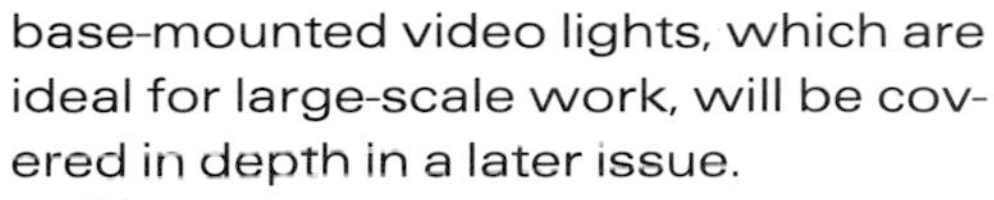

base-mounted video lights, which are ideal for large-scale work, will be covered in depth in a later issue.

Direction and intensity are also important considerations when choosing a battery lamp. Some lamps are hinged at the base so that you can angle the beam up and down. To concentrate the beam on a particular spot, purchase barn doors (flaps that fit to all four sides of the light and hinge in and out) for certain models. These allow you to control the spread and height of the light beam, so that you can direct it precisely on to your subject. Other lights are fitted with a zoom control, which enables you to narrow the beam on to a specific subject. This is especially handy when you are shooting scenes at the telephoto end of the lens.

POWER PROVIDERS

Video lights are powered in three ways. Some take power from the camcorder's own battery via the shoe, some from a Nicad battery, and

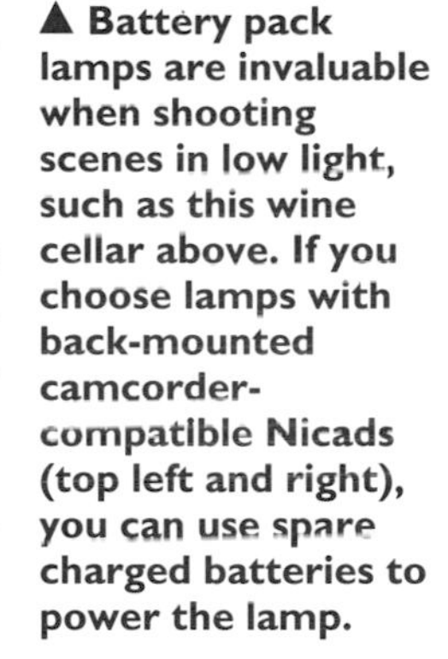

▲ Battery pack lamps are invaluable when shooting scenes in low light, such as this wine cellar above. If you choose lamps with back-mounted camcorder-compatible Nicads (top left and right), you can use spare charged batteries to power the lamp.

BATTERY CHARGERS

Most battery lamps come complete with their own plug-in charger. The model below – the Vanguard VL-503N – has a 6v battery which slides into a housing at the top of the lamp. This takes four hours to charge and will supply 20 minutes of light. For extra shooting time, purchase spare batteries with the lamp.

others have their own clip-on or built-in battery. The latter usually come with their own battery recharger.

Lamps tend to use up valuable battery time extremely quickly, so make sure you have plenty of spare charged batteries if you plan to shoot for any length of time.

Looking after your lamps

All lamps get hot. And the longer they are on, and the greater their power (more watts), the hotter they will get. Some of the more powerful clip-on models get hot enough to melt the plastic casing of small lightweight camcorders if left on for any length of time, and all lamps get hot enough to cause discomfort if they are touched, so take care not to burn yourself when shooting, and when removing the lamp from its fitting.

The most expensive replacement part of the video light will be the quartz bulb. These only last about 60 to 100 hours, and they are extremely delicate, so must be treated with special care. Their life is considerably shortened if the bulb is knocked or shaken when it is hot, so remember to allow the lamp cool down properly before moving it.

Quartz bulbs also need special handling. Never touch the bulb directly with your fingers when replacing it – any grease or residue will, at worst, make them explode when turned on, and, at best, shorten their lives. Handle them with thin cotton gloves or use a clean handkerchief to insert them. Once a quartz bulb has failed, replace the fuse as a safety precaution; even if it is still intact, it will have been stressed. □

CHECK IT OUT

- Beam power – ask to try the lamp in the shop when you buy. Some are badly balanced, with an intense, over-bright beam at the centre, fading at the edges. Choose one with a clear, even distribution of light.
- Colour balance – the colour temperature of video lights makes them ideal for indoor shooting, but you will get an orange tinge to the illuminated area if they are used as a fill-in light outdoors during the day. Ask your dealer about a blue daylight filter to fit over the lamp when you buy.
- Batteries – do you need any spares? Lamps consume a vast amount of power, so now is the time to buy.

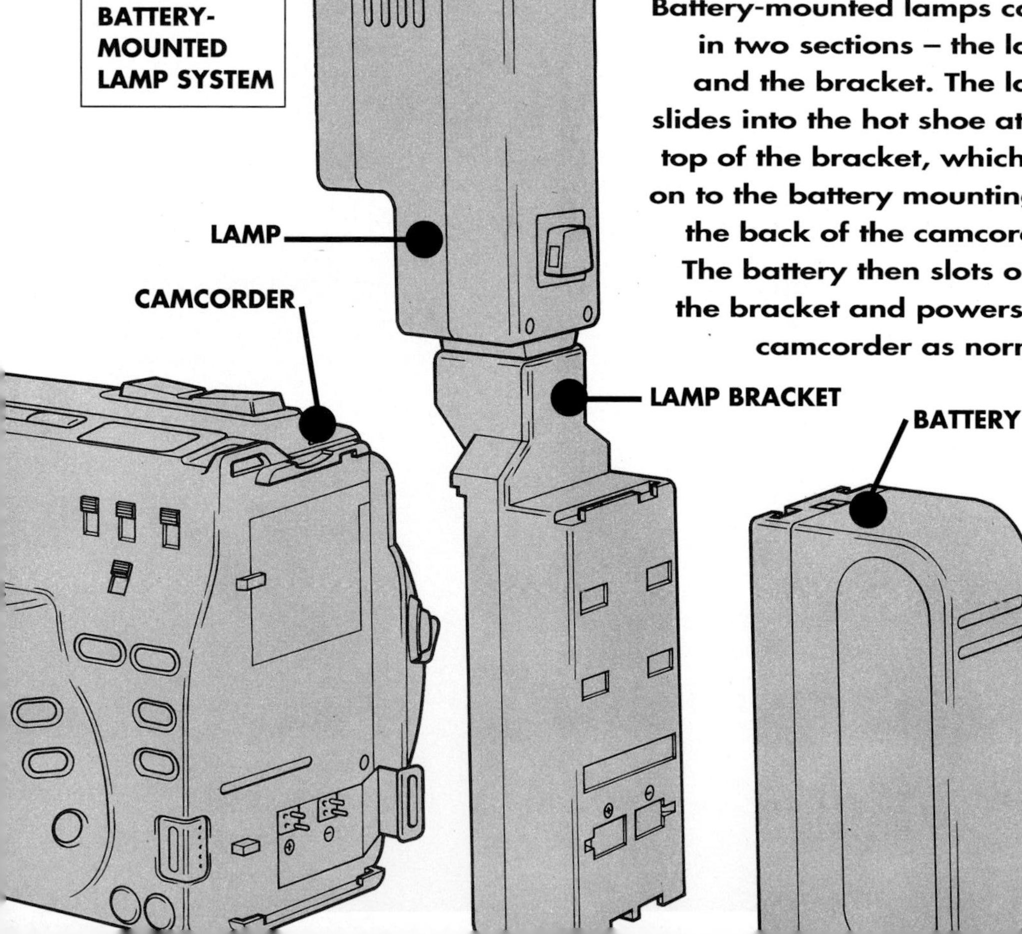

Battery-mounted lamps come in two sections – the lamp and the bracket. The lamp slides into the hot shoe at the top of the bracket, which fits on to the battery mounting at the back of the camcorder. The battery then slots on to the bracket and powers the camcorder as normal.

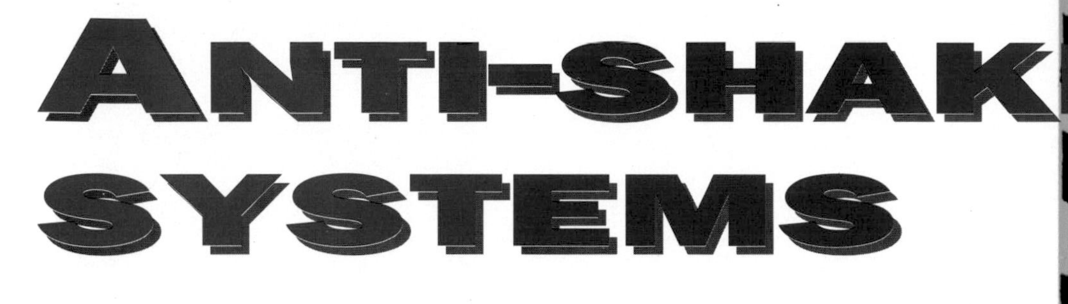

ANTI-SHAKE SYSTEMS

A ONCE-AND-FOR-ALL CURE FOR WOBBLY VIDEOS OR JUST ANOTHER CAMCORDER GIMMICK? WE TAKE A CONSIDERED LOOK AT PICTURE-STEADYING DEVICES

One of the greatest disappointments to any home videomaker is the inherent 'wobble' which always seems to be visible at some time other when the tape is played back. This is particularly apparent when using the zoom lens at telephoto settings.

This wobble or shake is caused by small involuntary movements of the hand – it is actually very difficult to hold a camcorder rock-steady, even with a two-handed grip.

Small and compact camcorders are convenient and popular with the buying public, but this trend to ever increasing miniaturization has also accentuated the shake problem. In order to compensate for this disturbance a number of manufacturers have introduced anti-shake systems on their smallest products.

TWO SYSTEMS

There are two sorts of system – electro-mechanical and digital – but both serve the same purpose. This is to sense those small involuntary movements and actually re-align the image in some way to compensate so that the actual on-screen picture seems rock-steady. What they won't do is compensate for big jumps and wobbles that might be caused, for example, by walking along with the camera, or shooting from a car. They are at their most effective in those situations where you cannot get firm support for the camcorder, especially when you are using the lens at its maximum telephoto setting.

The principle of such systems is simple. Sensors detect movement of the camera, this information is processed, and one of two things can happen depending on the system.

In an electro-mechanical system the optical path of the image through the lens is moved by a pair of prisms in the opposite direction to the movement of the camera so that the image falling on the pick-up chip remains stationary. In a digital system, there are no moving parts. Camera movement is either detected by sensors, or by comparing successive picture frames in a computer processor. This information is then used to move the sensitive area of the pick-up chip by re-assigning the individual pixels which make up the actual picture-sensing area. This 'movement' is in an exactly opposite direction to the accidental movement detected, so again the image appears stationary.

Each system has its advantages and disadvantages. The chief advantage of the electro-mechanical

ILLUSTRATIONS BY DAVID ASHFORD

JO BOURNE

▲ A hand-held shot on wide angle – perfect until you zoom in for a telephoto shot of the poppies. An anti-shake device would cure this.

system is the fact that there is no image distortion or degradation. The chief disadvantages are that it adds weight to the camcorder and uses a significant amount of power. As a mechanical system it could be prone to mechanical failure, although there is no reason to suppose that this particularly likely.

For the digital systems, no moving parts means that it is inherently faster at reacting to movement and is more reliable. Because it is chiefly electronic, it uses very little extra power, weighs very little and can also be connected to other automatic systems in the camcorder such as Autofocus and Auto-exposure to optimise performance.

But there is also a significant disadvantage. In order to allow the electronics to 'move' the sensitive area of the chip around, the system will only work on that part of the image which is in the centre 90 per cent or so of the frame. Some models blank out the part of the image around the edges they cannot process with a border, others enlarge that 'active area' digitally so that it fills the frame. Whichever method is used, there is inevitably some image quality loss. Further losses are caused because the high-speed shutter is also switched in. This is necessary to give the system time to analyse the image between successive frames and to make the necessary compensation adjustments.

Sensible shooting

The important thing to note about anti-shake systems is that they are not a substitute for sensible shooting practice. They will not give you the same steady image you would get by using a tripod or camera-brace and are not meant to be a substitute. They won't solve the wobbles caused by carelessly waving your camcorder about, so a tight, firm grip plus smooth movement from the hips for pans or tilts is still essential. What they are particularly good at is removing the involuntary shake and wobble, once you have a firm grip. Remember that and you won't be disappointed.

What's it worth?

Evaluating the value for money of such a system is another matter entirely, as you don't tend to see equivalent models both with and without the system. If you want to find out which systems are fitted to which models, consult your dealer.

The systems available are only effective in limited circumstances and many users will consider other camcorder features more important. However, if you are going to be doing a lot of shooting using the telephoto setting on the zoom lens or a lens extender in circumstances where a tripod is not a practical proposition, an anti-shake feature will prove to be very valuable. Otherwise it should be considered as useful, but only as part of a whole package of other features. □

Electro-mechanical System

Glass A compensates for vertical shake

Glass B compensates for horizontal shake

Liquid which has the same degree of refraction as glass

When the two pieces of glass are parallel, light goes straight through

When glass is not parallel, light refracts and image falling on pick-up chip remains stationary

Glass moves freely between the gap

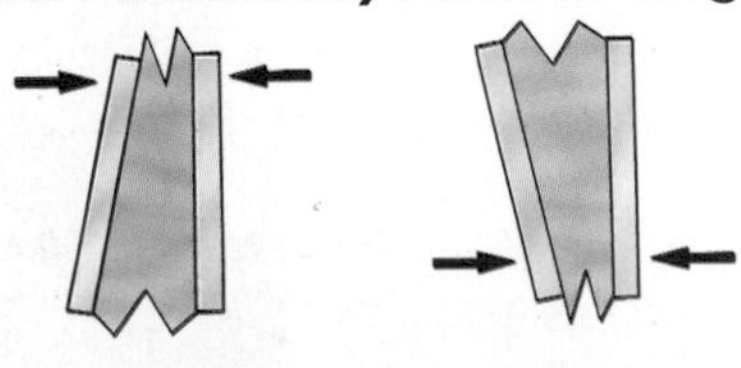

DIAGRAM BY MARK FRANKLIN

◀ Sony's electro-mechanical system. Two fine sheets of glass are fitted behind the camcorder's lens with a special liquid between. When the camcorder is steady, the sheets are parallel and light goes straight through. When the camcorder shakes, the liquid enables the sheets to move freely between the gap, light passing through refracts, and the image falling on the pick-up chip remains stationary.